Sociology of World Heritage

Taking mainly Japanese and other Asian case studies as examples, Ogino examines the motivations behind the preservation of objects and sites considered to be of cultural significance.

Using mainly the perspectives of Japanese approaches to cultural heritage, the book critiques the European logic of cultural heritage enshrined by UNESCO. It contrasts a Western emphasis on monuments and sites, with an Asian emphasis on more intangible forms of heritage, which place less emphasis on a linear view of time. More practically, the authors also analyze the positive and negative impacts that UNESCO-listed status has had on sites in Asia, including Angkor Wat, Nagasaki, and Lijiang. Finally, they address fundamental questions about who gets to decide what counts as cultural heritage, and what the underlying rationale is for actively preserving heritage in the first place.

This book is a thoughtful and provocative analysis of issues that will be of interest to sociologists, as well as scholars and students of cultural heritage.

Masahiro Ogino is Professor of Sociology at Kwansei Gakuin University, Nishinomiya, Japan.

Routledge Advances in Sociology

322 **Domestic Economic Abuse**
The Violence of Money
Supriya Singh

323 **The Paradigm of Social Interaction**
Nikolai Genov

324 **Identifying and Managing Risk at Work**
Emerging Issues in the Context of Globalisation
Edited by Chris L. Peterson

325 **The New Sociology of Ageing**
Martin Slattery

326 **The Class Structure of Capitalist Societies, Volume 2**
Social Space and Symbolic Domination in Three Nations
Will Atkinson

327 **We, Other Utopians**
Recombinant DNA, Genome Editing, and Artificial Life
Eva Šlesingerová

328 **Sociology of World Heritage**
An Asian Perspective
Masahiro Ogino

329 **Paramilitary Groups and the State under Globalization**
Political Violence, Elites, and Security
Edited by Jasmin Hristov, Jeb Sprague and Aaron Tauss

330 **The Disfigured Face in American Literature, Film, and Television**
Cornelia Klecker and Gudrun M. Grabher

For more information about this series, please visit: *www.routledge.com/Routledge-Advances-in-Sociology/book-series/SE0511*

Sociology of World Heritage

An Asian Perspective

Masahiro Ogino

Routledge
Taylor & Francis Group

LONDON AND NEW YORK

First published 2022
by Routledge
2 Park Square, Milton Park, Abingdon, Oxon OX14 4RN

and by Routledge
605 Third Avenue, New York, NY 10158

Routledge is an imprint of the Taylor & Francis Group, an Informa business

British Library Cataloguing-in-Publication Data
A catalogue record for this book is available from the British Library

Library of Congress Cataloguing-in-Publication Data
Names: Ogino, Masahiro, 1957– author.
Title: Sociology of world heritage : an Asian perspective / Ogino Masahiro.
Description: New York : Routledge, 2021. | Series: Routledge advances in sociology | Includes bibliographical references and index.
Identifiers: LCCN 2021029244 (print) | LCCN 2021029245 (ebook) | ISBN 9780367857608 (hardback) | ISBN 9781032138152 (paperback) | ISBN 9781003014904 (ebook)
Subjects: LCSH: World Heritage areas–Social aspects–Asia, East. | Cultural property–Social aspects.
Classification: LCC G140.5 .O45 2021 (print) | LCC G140.5 (ebook) | DDC 363.6/9095–dc23
LC record available at https://lccn.loc.gov/2021029244
LC ebook record available at https://lccn.loc.gov/2021029245

ISBN: 978-0-367-85760-8 (hbk)
ISBN: 978-1-032-13815-2 (pbk)
ISBN: 978-1-003-01490-4 (ebk)

DOI: 10.4324/9781003014904

Typeset in Galliard
by Apex CoVantage, LLC

Contents

List of figures   vi
Preface   viii

Introduction   1

1   The birth of museological desire   9

2   War and world heritage   22

3   Semantics of inscription for the World Heritage List   48

4   Intangibles and tangibles: The logic of actualization   72

5   Age of preservation   94

6   Frozen time–space and cultural heritage   118

References   140
Index   146

Figures

0.1 Reproduction of a housing complex at Matsudo Museum to the
     left, real housing complex to the right   4
2.1 Tourists climb to the summit of Angkor Wat   24
2.2 The Banteay Srey Community Tourism   25
2.3 Killing fields   26
2.4 Tourists around the Great Onofrio Fountain in Dubrovnik   27
2.5 Genbaku Dome   30
2.6 Memorial Cenotaph for the A-Bomb Victims and in the
     background, Genbaku Dome   31
2.7 Japanese sandal and bag used by Sadako during her
     hospitalization exhibited at the Hiroshima Peace Museum.
     Donation by Shigeo Sasaki and Masahiro Sasaki   33
2.8 High-school students and volunteer guide at the hypocenter   36
2.9 Drawings and photographs exhibited at Hiroshima Peace
     Memorial Museum. The drawings by Eichi Ueda and Yoshio
     Takahara. The photographs by Masami Onuka and Japanese
     Army Ship Command are part of the collection of Shogo Nagaoka   38
2.10 The city hall of Bikini Atoll removed to Majuro Atoll   45
3.1 Trends in the criteria for selection of World Heritage sites. The
     statistics and its graphic design was done by Mayumi Yukimura   50
3.2 Tajima Yahei's House   52
3.3 Invisible Ebisugahana Shipyard   54
3.4 A group of tourists go to the Reverberatory Furnace from the
     parking lot   55
3.5 A drawing shows the monument in its original state on the
     panel at the remnant of Ohitayama Tatara Iron Works. A certain
     imagination is required to "see" the site   55
3.6 A former factory building of the Mitsubishi Shipyard today used
     for the museum of the shipyard, a component of the World
     Heritage site   58
3.7 Hashima Island   59
3.8 Yunotsu town   63
3.9 Gorin village and Former Gorin Church in Kuga Island   64
3.10 Mount Fuji   68

3.11 Miho no Matsubara without Mount Fuji not visible on a crowdy day 70
4.1   Tôdaiji 76
4.2   Naikû (Inner Shrine) of Ise Shrine. Pictures can only
      be taken outside 77
4.3   Gekû (outer Shrine) and *kodenchi*, empty place reserved for the
      reconstruction 78
4.4   Empty place for ritual practices in Ise Shrines 79
4.5   A House of *Oshi*, house for pilgrims 80
4.6   Evolution of the rating of the three artists (1985–1994)
      (Ogino, 2002:214) 88
4.7   Awaji Ningyô Za (Awaji Puppet Theatre) 89
4.8   Ashio Copper Refinery out of use 91
5.1   Traboules in Lyon 97
5.2   Mailbox for residents in Traboules 98
5.3   Medina in Tunis 99
5.4   A poster of classic music concert at Carthage in French
      language in Tunisia 99
5.5   El Jem in Tunisia, preparation of a concert 100
5.6   Aposter shows the removal the Dresden Elbe from the World
      Heritage List with the sentence "Anger and tears remain" 101
5.7   The *Waldschlößchenbrücke* in Dresden 102
5.8   Tourists in nightlife of Lijiang, China 106
5.9   Tourists go to the Mont-Saint-Michel 106
5.10 798 Art District in Beijing 108
5.11 Former ceramic factory in Jinderzhen, China 108
5.12 "Dentist" by Shinro Ohtake in Naoshima 109
5.13 Active Fault preserved in Awaji Island 112
5.14 *Kapal di atas rumah*, The Boat on the roof after the great
      tsunami in Banda Aceh, Indonesia 113
6.1   The old shopping street in Takasago city 129
6.2   A street in Tatsuno city designated as Important Preservation
      District of Traditional Buildings 135
6.3   Himeji Castle and main street in Himeji city 137

Preface

I started to study, from a socio-anthropological point of view, the issue of cultural heritage when Henri-Pierre Jeudy asked me in 1994 to write an article for the journal *Ethnologie Française*, on heritage in Japan, in particular, on intangible cultural heritage, a specifically Japanese notion. The continuous discussion with Henri-Pierre Jeudy paved the way for a larger Japanese-French collective research on the issue with two other French researchers, François Séguret and Marc Abélès, and three Japanese researchers, Kenichi Wakita, Nobuhiko Ogawa, and Yoshiyuki Yama, from 1997 to 2000. We organized a conference at the Ecole des Hautes Etudes en Sciences Sociales and published research results as a collective book in Japanese in 2002. This book has unexpectedly echoed in the sociological community and other disciplines in Japan. The issue of intangible cultural heritage has also attracted the attention of Chinese scholars, and the China Society for Anthropology of Art has invited me to give a lecture several times since 2010. The participation in the workshop and conference "Cultural Heritage? in East Asia", organized in 2010 by the Sainsbury Institute for Study of Japanese Arts and Cultures, in which Akira Matsuda and Luisa Elena Mengoni played a crucial role, allowed me to deepen the reflection on the question.

The rise of these interests toward the question of cultural heritage is closely related to a global transformation. I had already alluded, in the first article in French, to the underlying relationship among cultural heritage institutions, the nation-state, and the market economy. But I increasingly felt the need to take into account the global dimension of the issue with the growing influence of World Heritage institutions. That is why I undertook, together with Mayumi Yukimura, the research on the relationship between the globalizing effects of World Heritage institutions in different societies, especially in Japanese society.

Having said that, and despite the subtitle given to this book, that is, Asian Perspective, it is not intended to make a naive culturalist contrast between the East and West. If this book focuses mainly on Japan and non-European countries, it is because these societies show more explicitly the effects brought about by World Heritage institutions. However, this does not prevent us from seeing the same effects in European societies, in which I also conducted the survey.

This global transformation allows us to reflect on two issues. The first is the question of the relationship between cultural heritage institutions, the state, and the capitalist system. The capitalist system is, by its nature, globalizing, but this characteristic has been accelerating since the 1980s. Curiously, World Heritage institutions were implemented at the same time. The national cultural heritage institutions were then influenced by it. This globalization, both economic and cultural, and the national and nationalist resistance to it, should be a major focus of sociological research. Second, this sociological question must be analyzed in terms of time and space, as some social scientists point out through other research fields. I would be very happy if this book could give rise to further reflection on the question of time–space.

I am grateful to Eric Hsu, who was kind enough to read the manuscript carefully and give a helpful commentary. I thank Fang Lili and Wang Yongjian, who invited me in 2016 as an expert to give a seminar at Chinese National Academy of Arts on this issue, and Ralf Futselaar, who organized a seminar for us at The Netherlands Institute for War Documentation in 2013. I also thank Shôichirô Takezawa, who invited me to participate in the collective research on "negative" heritage at the Japanese National Museum of Ethnology and Yang Fuquan and Li Yongxiang for my survey in Yunnan province in China and all the staff and officials of the world heritage organizations whose names I will not mention. This research was supported by the Grant-in-Aid for Scientific Research of Japanese Ministry of Education, Culture, Sports Science and Technology (2013–2015) and the Special Fund of Kwansei Gakuin University (2016).

Introduction

The paintings that can be seen inside the Pantheon show Paris, the eternal city, several times invaded, but each time saved from the invaders' hand. These paintings symbolically summarize my first impression during the visit in the 1970s: this city is governed by a relentless desire for conservation. Nevertheless, the number one enemy of Paris is not the foreigner, but simply time. Indeed, there is an immense effort to preserve historical objects and monuments against the wear and tear of time.

Monuments, by their very presence, ensure the continuity of history. The inhabitants are thus led to recognize a linear conception of history. Objects that are not immediately accessible to the public are transported to museums that give them historical value. By no longer having their original use, they simply embody the past in the museum setting.

> Museums are first and foremost made for objects. They do not belong to us. They come from far away in time, we have received them from previous generations and our first duty is to pass them on intact to those who will come after us.
>
> (Chiva and Lévi-Strauss, 1992:170)

These words of Claude Lévi-Strauss reflect the nature of uprooted objects, which serve to show the duration and continuity of history and the function of the museum as an institutional framework for the preservation of these objects. The objects exhibited in museums are definitively separated from everyday life. After the French Revolution, the palace itself became the museum known as the Louvre. At the same time, the notion of historical monument appeared. The preserved historical monuments are like the objects exhibited in the museum.

This desire to preserve objects and monuments can be called museological desire. Museological desire developed in France from the revolution. Other European countries followed the path blazed by France, and the Louvre became the model of the art museum (Duncan, 1995:32).

Raymond Williams pointed out that the words culture and art appeared in the second half of the eighteenth century along with words such as industry, democracy, class (Williams, 1960). Culture and art would have played a crucial role in the constitution of the nation-state, and since then, the cultural and political

DOI: 10.4324/9781003014904-1

spheres have been closely linked. To clarify the relationship between these two spheres, Louis Althusser put forward the idea of the ideological apparatus of the state, of which school and family are part (Althusser, 1976). To this can be added the cultural heritage institutions.

Museological desire appears at a specific time; not all societies have museological desire. In fact, this was not the case for Japan. The national museums did not play a part in cultural authority as in Europe. The Museum Promotion Association was established in 1928, and President Shigenobu Hirayama submitted a proposal to the Minister of Education to improve museum facilities. In 1929, when the Law for the Preservation of National Treasures was enacted, several individuals who were involved in the museum sector (nine individuals from the Imperial Museum, the Ministry of Education's Religion Bureau, the Tokyo Fine Arts School, and the Museum Promotion Association) gathered to discuss issues pertaining to the preservation of cultural heritage. One of the key issues they addressed was how to collect elements of Japan's cultural heritage at a national museum. The following exchange reveals the state of cultural heritage protection at that time:

- Yashiro (Professor, Tokyo Fine Arts School): So how effective will it be to order people to send national treasures to a museum?
- Oshima (Director, Imperial Museum): There's no way we can force people to send things in by issuing an order.
- Yashiro: Even if we were to try, they wouldn't take it very well.
- Oshima: That's why I think that we need to win their understanding without issuing an order (Yashiro, Oshima et al., 1929).

This exchange shows how difficult it is for museums to collect items even when a law has been enacted to facilitate the collection of cultural heritage objects. Moreover, Inô Dan lamented the lack of curators in Japan (Dan, 1928). Without specialists taking care of the functioning of museums, they could not function well. Interest in cultural heritage preservation temporarily appeared under the leadership of GHQ Supreme Commander Douglas MacArthur after World War II, and repair work was performed on Kyoto's abandoned and neglected Nijo-jo Castle, but efforts to preserve cultural heritage objects did not begin in earnest until the mid-1960s. This contrast between Western Europe and Japan is not simply a cultural difference. It is a question of how a society is organized. Or more precisely, it is related to the way a society constitutes modern state institutions.

The situation changed radically from the moment the World Heritage institutions were established in 1972 and introduced new rules for the preservation of cultural heritage in countries that ratified the World Heritage Convention. There are, at least, two elements of change brought by the World Heritage institutions. First, the issue of cultural heritage is no longer limited to the interior of a nation. It becomes local, national, and global at the same time. Second, the transformation is not limited to the sphere of heritage preservation itself. Indeed, the World Heritage institutions contribute to bringing about wider change throughout society and transform the relationship between economic, political, and cultural structures within a society. This change is defined as the coming of the Age of Preservation. In this age, heritage preservation becomes more valued, and the nature of those recognized as cultural heritage is more diverse. In fact, increasingly, what are saved are more than just cultural heritage recognized as having aesthetic value, or assets from the distant past such as ancient Egyptian mummies.

The Age of Preservation brings three changes to cultural heritage institutions themselves and to the way a society values objects and monuments. Three characteristics can be identified, all of which contribute to the removal of existing boundaries. First, the Age of Preservation removes the boundaries between the past and the present. While, in the classical conception, cultural heritage must be part of a distant past, the time that it takes for a monument to become a heritage is increasingly reduced. There are a number of cultural heritage sites of the past that is so recent that people feel it to be almost a part of the present. One of first illustrated examples is Brasilia. The city was constructed from 1956 to 1960, and only 27 years after the creation of the city, Brasilia was inscribed for the World Heritage List. Brasilia is not an ancient vestige. It is a futuristic city and a rare faithful realization of modern urbanism. The inscription of Brasilia brought, at least, two notable changes. First, the time elapsed since a monument was constructed is less important; Brasilia is valuable, because it is modern. Second, Brasilia functions actively as the capital of Brazil; it is not a cultural heritage of an old type that can be seen only from afar. In this sense, Brasilia as World Heritage is in the present, and the present of the city itself should be preserved as a whole. The preservation of the present also applies to the monuments like the Genbaku Dome (A-Bomb Dome) that invoke abhorrent and still vivid memories and are preserved even if some people may want to quickly expunge them.

It even happens that the inhabited space simultaneously becomes an object exposed in the museum. This is the case of a housing complex, Tokiwadaira, in Japan. The Matsudo Municipal Museum built, in the museum, a reproduction of a part of a housing complex Tokiwadaira built in 1961 based on the original design, while right next to the museum, there exist the buildings still inhabited (Figure 0.1). The museum also reconstructed the interior of the apartment equipped with what was then used as a TV set at that time and so on. It was around this time that similar housing complexes began to be built in suburban areas of Japan. Two bedrooms with an eat-in kitchen is a typical model for apartment complexes. The housing complex is thus turned into a self-referential heritage and constitutes a strange scene in which the present is already on display in the museum.

Second, the Age of Preservation takes away existing barriers between the inside and outside of cultural heritage sites, as the example of Brasilia shows well. Traditionally, there was the barrier between a cultural heritage site and the rest, as Benedict Anderson pointed out as following.

> The grandeurs of Borobdour, of Anghor, of Pagan, and of other ancient sites were successively disinterred, unjungled, measured, photographed, reconstructed, fenced off, analysed, and displayed.
>
> (Anderson, 1991:179)

Figure 0.1 Reproduction of a housing complex at Matsudo Museum to the left, real housing complex to the right

Source: Photo by the author.

Thereafter,

> the reconstructed monuments often had smartly laid-out lawns around them, and always explanatory tablets, complete with datings, planted here and there. Moreover, they are kept empty of people, except for perambulatory tourists (no religious ceremonies or pilgrimages, so far as possible).
>
> (Anderson, 1991:182)

At the Age of Preservation, contrary to what evoked Anderson, monuments are not kept empty of people. On the contrary, on and around many World Heritage sites, the inhabitants continue to live. Most of the examples in this book show the removal of barriers that previously protected cultural heritage sites.

Finally, the Age of Preservation tends to eliminate the distinction between those who give cultural value to objects and monuments, and others. In the first phase, cultural heritage institutions created the unequal power relations between promoters of heritage planning and those who accept this planning, such as the relations between the colonizers and the colonized. But in the second phase, local people are beginning to actively discover what is likely to be a cultural heritage. Those who promote heritage creation invent various strategies to carry out their project. Local governments and communities make efforts to have their heritage inscribed onto the World Heritage List. Furthermore, World Heritage sites are transforming the way we see and approach the world. The development of tourism is accelerating this change. Indeed, World Heritage sites can attract tourists who, during their travels, have the experience of visiting another world and being strangers themselves. When people experience such transitional places like tourist destinations – which are by definition outside their everyday lives – they also start seeing the world to which they return as something external. As for residents living within or near a World Heritage site, they are forced to live with tourists every day. They have no choice but to accept outsiders. Therefore, they begin to share the point of view of tourists who meet a world for the first time and end up seeing the world as something external, and moreover they begin to behave in accordance with previously external views and desires. Tourists and residents living near a world heritage site, therefore, both live in an unstable situation in which they feel they are in two places at once – the world they live in and the world they visit. People seem to drift back and forth between where one lives and external destinations. This situation can be called the doubling of the world (Ogino, 2016:25). World Heritage institutions, thus, imply the change of power relations.

The Age of Preservation is observed from the 1980s onwards and is linked to the global change of the capitalist system. The capitalist system seeks to produce and sell items as quickly as possible to make more profit. Speed becomes important and hence the duration is considered negative for production. This tendency is accentuated since the 1980s. According to Zygmunt Bauman, being transient becomes a value and duration is less important (Bauman, 2005). In the same way, David Harvey designates the ephemerality as the dominant notion of time in consumer society (Harvey, 1989). Speed becomes important, and hence the duration is considered negative for production. This acceleration of production goes hand in hand with the expansion of the consumer society. In consumer society, consumption is no longer an act of buying just what they need and becomes an activity that is itself cultural. Fashion is a key word in the consumer society. It shows explicitly the development of ephemeral time.

Curiously, as ephemeral time dominates, interest in cultural heritage increases. This phenomenon becomes an object of research and is first interpreted in terms of collective memory. Barbara Misztal asserts that the concept of collective memory has become one of the more important topics addressed in today's social sciences (Misztal, 2003:1). The notion of collective memory is, thus, now coming to the forefront in the social sciences and humanities. This concept is first defined by Maurice Halbwachs (Halbwachs, 1925). For Halbwachs, individual memory is not completely closed insofar as it is constructed through interactions with others. Individual memory is, then, influenced by the collective representation of the past. He also notes collective memory is essentially a reconstruction of the past. He goes so far as to say that knowledge of what was original is secondary. Collective memory considered in this way comes close to collective representation. It is more than a simple aggregation of individual memories.[1]

The historian Pierre Nora has taken up this notion but has only taken up its second aspect, that is to say, collective memory almost synonymous with collective representation. Nora notes that modern society is losing its sense of memory and needs to actively set up symbolic elements that will allow society to retain its memories. He designates these elements as *lieux de mémoire*, that is, sites of memory (Nora, 1984). Some, like museums or monuments developed in modern society, play an important role in the reconstruction of collective memory. Public institutions like archives, libraries, and museums have thus been increasingly established since the nineteenth century. Important events are commemorated, anniversaries are celebrated, and monuments are constructed to remember these events. Autobiographies also offer *lieux de mémoire* that historians can explore.

Almost at the same time, Henri-Pierre Jeudy publishes *Mémoire du social* (Memories of the Social) (Jeudy, 1986), practically the first study in sociology about cultural heritage institutions.[2] Historians might say such or such monuments have historical value, where art historians might say certain monuments have aesthetic value. But sociologists must ask themselves other types of questions – why does society preserve its cultural heritage? How have we come to think that ancient monuments are worth being preserved? He tried to answer these questions by showing that the growing interest in collective memory and cultural heritage is closely linked to the question of death. Since the publication of Philippe Ariès' work (Ariès, 1975), historians and sociologists have advanced the idea that relations with the dead are increasingly limited. He contradicts this thesis by showing that cultural heritage institutions induce a way of living daily relationships with the dead (Jeudy, 1986:11).

Nevertheless, the concept of memory is not sufficient to analyze the emergence of the Age of Preservation. The question of space must also be taken into account. In fact, Halbwachs links memory to a place when he analyzes the place of pilgrimage (Halbwachs, 1941). Nora's term, site of memory, also signifies that the memory is closely related to the question of space. It is therefore necessary to relate memory or time to space. If we consider memory, we also have to consider space. In particular, for the stability of society, space plays a primordial role. When our intimate space is unchanged, we feel secure. When we see the same landscape every day, our life seems stable. The stability of space is, thus, a condition *sine qua non* to show the continuity of the society. The stability and continuity of space can be defined as "identity of space".

Capitalist development jeopardizes the identity of space on several levels. In the first place, the absence of duration and the extension of temporal fluidity implied by the consumer society creates the ambiance of instability. In addition, the process of industrial development necessarily involves the transformation of space and sometimes creates the destructured landscape.[3] It is the landscape in which there are two opposite types of landscape. One is traditional, the other modern. The destructured landscape shows a chronological disorder and suffers the loss of identity of space because of rapid development. To improve this situation, cultural heritage institutions are mobilized nowadays. In other words, the contradiction between the economic and the social emerges in a new dimension today, and cultural heritage institutions become a possible remedy. But they are not always effective in resolving this contradiction.[4] And in order to grasp these intertwined relations between the economic, social, and cultural, a new theoretical dimension must be introduced. This is precisely the dimension of time/space. Anthony Giddens remarked that neither time nor space has been incorporated into the center of social theory (Giddens, 1979:202). Since then, the situation has not changed much,[5] and the time–space relationship has not been sufficiently taken into account. Instead of treating space and time separately, we, then, need to consider the time–space relationship. This attempt could bring an innovative sociological understanding.

Chapter 1 will explain the birth of cultural heritage institutions using the concept of museological desire versus capitalist desire. These institutions are driven by museological desire, which did not exist before the development in the nineteenth century of capitalism and also of the nation-state in Western Europe. This chapter will first explain why museological desire appeared precisely at that time. Then, the second part of the chapter will show the transformation of cultural heritage institutions after World War II, which led to the creation of World Heritage institutions in 1972.

The creation of World Heritage institutions is, in a sense, the result of World War II, which was the consequence of the struggle between the nation-states. The 1954 Hague Convention notes that cultural property was severely damaged by the war and that protective measures were needed. UNESCO followed up on this issue and the World Heritage convention was formed. The institutions created on this occasion played an important role not only for the protection of heritage but also for the recovery of the society affected by the war. On the other hand, the tragedy caused by the war itself has become an object of World Heritage, which is called in Japan "negative heritage". These issues on the close relationship between war and world heritage are discussed in Chapter 2.

Chapter 3 shows the transformation in the conception of cultural heritage brought about by the establishment of World Heritage institutions. Indeed, traditional standards of cultural heritage protection can be difficult to uphold in certain societies. Sites affected by bombing must be restored, and restored monuments are considered to lose authenticity and integrity, two classic conditions that give value to cultural heritage. In addition, there are societies that do not have the same conception of cultural heritage. Wooden buildings wear out more easily than stone buildings. It is therefore difficult to conserve wooden buildings without constantly restoring them. The principle of authenticity and integrity from the European culture cannot be applied to all societies. Thus, in the Nara Document, adopted on the occasion of Nara Conference held in 1994, cultural diversity and respect for the cultural tradition of all societies are emphasized. From this point on, the representation of a site becomes more important because the tradition itself is intangible and must be represented. Therefore, building up a narrative that a site may contain is required for the inscription on the World Heritage List, and certain semantics are constituted. By taking Japanese cases, Chapter 3 then will show characteristics of these semantics.

Chapter 4 will deal with the intangible aspect of cultural heritage that Nara Document underlines again by taking the Japanese example. In Japan, this intangible aspect is particularly accentuated. The notion of legally defined "intangible cultural property" illustrates this fact well. This aspect is related to the way in which the nation-state was constituted, which is different from that of Western Europe. Chapter 4 will explain this through the notion of "the logic of actualization". This logic consists in actualizing in the present what once existed or is believed to have existed in the past instead of faithfully preserving the heritage of the past. The existence of this logic shows that the way of preserving cultural heritage varies from one society to another. But with the arrival of the Age of Preservation to Japanese society in the 1980s, the conception of cultural heritage has changed. That will be pointed out in the last part of this chapter.

Chapter 5 will, first of all, detail the various aspects of the Age of Preservation evoked in this Introduction. World Heritage sites are largely affected by the Age of Preservation. There are cases where sites adapt well, and there are cases where they do not. Then, this chapter will show the relationship between the Age of Preservation and various aspects of social change as the development of the consumer society and radical change brought by natural and social disasters. And finally, Chapter 5 will point out that in the Age of Preservation, at the limit, everything can have the right to acquire cultural value. But if this is the case, we may ask ourselves if any of them have no value.

The last chapter, Chapter 6, will discuss about this risk of the annihilation of the value of cultural heritage using the time–space approach. One of the characteristics of the Age of Preservation is the disappearance of the distinction between the past and the present. Chapter 6 will then show how this feature came about. From the 1980s on, there is no longer a clear orientation toward the future, and sense of duration is increasingly lost. This absence of duration creates uncertainty and the role of space increases. It is at this point that the Age of Preservation appears. Given this theoretical reflection of the time–space relationship, the last chapter attempts to answer the question: for whom does cultural heritage exist, and should it be preserved?

Notes

1 Jeffrey Olick notes that this concept has two categories: socially framed individual memories, and collective commemorative representations and mnemonic traces (Olick, 2007:20).
2 A little later, Paul Connerton published *How societies Remember* (Connerton, 1989).
3 The great disaster also disrupts spatial stability. The identity of space is no longer assured when a catastrophe occurs. If nothing is done, the future of the area is completely uncertain. In this case, the landscape is completely destroyed.
4 The destructuration of the landscape becomes a problem even in cities that have World Heritage sites. A famous example is that of Dresden, a site that was deleted from the World Heritage list due to the decision to build a bridge there. UNESCO considered that a new bridge could destroy the balanced landscape approved by the World Heritage institutions. This issue will be analyzed in Chapter 5.
5 The sociology of time and the sociology of space have developed since then but each in an independent way. There has not really been an attempt to relate time and space theoretically.

1 The birth of museological desire

The appearance of museological desire, desire to preserve valuable objects an monuments should be explained in relation to the capitalist system and the nation-state. Indeed, it develops within the framework of the nation-state and at the same time as the expansion of the capitalist system. The relationship of museological desire to the nation-state is clearly illustrated by what happened during and after the French Revolution, beginning with the creation of the Louvre.

The Louvre project was advanced before the revolution. The king's director general of royal buildings, Comte d'Angivillier, had the plan for the creation of a museum in the Grand Gallery of the Louvre. "The Louvre was to be a source of national pride as well as royal glory" (Maclellan, 1994:49). The Louvre project, under the influence of the Enlightenment, appeared before the revolution and made art a pivot of the nation. An essential role had already been assigned to art to ensure the nation's continuity.

The Louvre project was realized after the revolution. Andrew McClellan quotes a letter from the Minister of the Interior, Jacques-Louis David, who states, in 1792, "the museum will become among the most powerful illustrations of the French Republic" (Maclellan, 1994:92). The revolutionary government wanted to make Paris the capital of art, like Athens in the past. The palace of Louis XVI became the Louvre and opened in 1793. Churches were also transformed into museums. In doing so, buildings, which had belonged to royal and aristocratic families opened to the public, and various objects started to be exhibited. The museum was considered an ideal of revolution (Poulot, 1997) and become a sacred place of art and incited a secular ritual. Visitors of the art museum visit the new sacred place like pilgrims. Carol Duncan rightly points out the decisive role played by the Louvre.

> As a new kind of public ceremonial space, the Louvre not only redefined the political identity of its visitors, it also assigned new meanings to the objects it played and qualified, obscured, or distorted old ones. Now presented as public property, they became the means through which a new relationship between the individual as citizen and the state as benefactor could be symbolically enacted.
>
> (Duncan, 1995:24)

DOI: 10.4324/9781003014904-2

The Louvre has made it possible to link the individual to the state through the objects on display. As individuals become citizens, visitors to the Louvre learn about the progress of society through the history of art. This progress is achieved by the republic. Moreover, this progress is measured in terms of the universal ideal of beauty (Duncan, 1995:25).

Order of recollection

What is it, then, that causes museological desire to develop in the nation-state? It is due to the expansion of the capitalist system, which has fundamentally endangered the existing social order. What then is this traditional order? According to Émile Durkheim, in an elementary society, "cohesion is essentially due to the community of beliefs and feelings" (Durkheim, 1978:261). And he continues, "what brings people together are mechanical causes . . . such as blood affinity, attachment to the same soil, ancestor worship, community of habits" (ibid: 262). "Blood affinity" and "ancestor worship" play, thus, an important role in the constitution of social order.

His nephew Marcel Mauss introduced a new dimension to the idea of Durkheim. He pointed out, in a passage from his research on the gift, "among the first groups of beings with whom men must have made contacts were the spirits of the dead and the gods" (Mauss, 1978:13).[1] Maussian indication means the primordial exchange is an exchange between the living, the gods, and the spirits of the dead, because we have a strong sense of "obligation" toward our ancestors. The order, based on the recollection of the dead, is therefore established through a never-ending exchange between the living and the spirits of their ancestors. We call this "the order of recollection" (Ogino, 1998, 2008a:59–61).

The order of recollection is the result of man's greatest concern – how to deal with the death of a loved one. In the order of recollection, order is sustained by their memories of the dead. The dead are seen as protecting the living, and ancestors can at times symbolically communicate with the living. As a result, this imbalance becomes a source of control, by demonstrating that one possesses special powers that enable communication with the gods and the spirits of the dead. For example, in Japanese Buddhist tradition, the spirits return to *shigan* (the human world) during the *obon* season and therefore the living carry out rituals to receive their ancestors, such as making daily offerings to the *butsudan* (Buddhist altar).

The same goes for Christianity. The Rector of the Basilica of Lisieux in France, dedicated to Saint Theresa, Raymond Zambelli, affirms that "our missing are present but with a presence hidden from us. They are active, but in their own way" (Zambelli, 1995). Especially, the rector emphasizes active presence and prodigious influence of Saint Theresa on the Earth for almost a century. Lisieux is therefore a recognized place of pilgrimage. Among pilgrims who visit the basilica, there are those who have some problem, for example those who are suffering from alcoholism.[2] They pray for recovery from alcoholic addiction. Their own efforts are not enough. The implicit presence of Saint Theresa helps alcoholics to overcome their problem. It is believed that only by recalling a sacred existence like Saint Theresa, they can escape from the addiction.

As the rector pointed it out, the ancestors are not here below. They are supposed to exist somewhere else. That said, people can symbolically interact with the ancestors only in certain places. According to Mircea Eliade, at first, people fixed a "sacred center" such as mountains inhabited by deities, temples, and palaces (Eliade, 1969:23). The center is seen as a place linked to the world inhabited by the ancestors and spirits. At religious capitals such as Mecca and the Vatican, the entire city also becomes a center continually visited by pilgrims. Lisieux is one of these cities that became places of pilgrimage.

Crisis and restoration of social order

The order of recollection constitutes a system for dealing with a crisis that could be a natural disaster such as a flood, or a deterioration of social order triggered by a crime such as theft. It attempts to restore order and overcome a crisis, by arriving at an understanding of the cause of the crisis by relating it to gods and spirits.

The Legends of Tono[3] (2008) and its supplement (1989), gathered by ethnologist Kunio Yanagita, show well the process in which a community solves problems through the order of recollection, insofar as the legends describe in detail what the habitants remember about the area of Tono and its habitants. Among the legends, take a typical anecdote of theft that demonstrates how the recollection of order functions.

> In the event of an adversity such as theft, the villagers call on the powers of a god named *mitsuminesama* in order to identify the perpetrator. A man called Yoshitaro Sasaki called on the powers of *mitsuminesama* when *watagase* (ceremonial dress made of cotton) was stolen from his house. All the lights in the house were turned off and the *goshintai* (body of god) placed in the *okuzashiki* (innermost room of the house). Each of the villagers who had gathered in the house then entered the room to pray to *mitsuminesama*. One woman, however, was terrified and refused to enter the room. When the villagers tried to force her into the room, the woman vomited blood and lost consciousness. The villagers therefore concluded that she was the thief. The woman handed over the stolen goods to the villagers that night.
>
> (Yanagita, 1989:101–102)

As is clear from this anecdote, the villagers believed that it was *mitsuminesama* and not they themselves who have the ability to solve crimes of theft. Therefore, a series of rituals were carried out in preparation for this, commencing with the welcoming of the *goshintai* by the villagers. This ritual is clearly only carried out "when there are already several suspects" (Yanagita, 1989:101) and the villagers already have some idea of who is guilty. It should actually be possible to identify the perpetrator without having to carry out such rituals. However, the act of reconfirming the function of the order of recollection through the ritual of identifying the perpetrator is of more significance to the villagers than simply identifying the perpetrator for the Sasaki family and returning the *watagase* to them. Thus, in the event of an incident, the villagers must ask the spirits of the dead for assistance.

Violence was indirectly exerted in this example of theft by forcing the female suspect to vomit blood through the act of ritual. The community exerts violence directly or indirectly against those who it believes have caused a deterioration of order. In each case, the violence is understood to be promoted by a god such as *mitsuminesama*. The exertion of violence is therefore validated because of a god or deity who inhabits a space found beyond the boundaries of the village.

We can see from the earlier section that the order of recollection actuates the device that in turn enables the exertion of violence in response to a crisis. If a priest or a priestess is supposed to have the ability to communicate with the gods, the villagers believe their oracle to be true because they are mediators who are able to communicate with the gods and the spirits of the dead. This signifies that the villagers are hierarchizing the world that they live in, including places accessible to any person through to those only accessible to exceptional people or beings. The ability to demonstrate one's power to communicate with gods becomes a source of control.

The order of recollection is also mobilized to face the natural disaster. It helps mitigate the grief of those who lost their loved ones and to symbolically overcome the difficulty. Temples constitute places of mourning those who died unexpectedly. If temples are damaged by a disaster, they should be immediately restored. The reconstruction of damaged temples allows the recreation of a center, intersection of the living and the dead. Their reconstruction is somehow an act of restoring social order. The order of recollection helps to console those who have lost their loved ones because of a disaster. In doing so, the order of recollection constitutes an understanding system. It helps understand the unexpected death of loved ones by affirming they exist somewhere else.

This understanding system also functions when a community accepts new objects from outside. In an example taken by Mauss, Trobrianders have to organize a ritual to do *kula*, a sort of gift trade. When they receive gifts from other tribes in *kula*, they recite a long enchantment formula. In the formula, they recite they have to ward off all the feeling of hatred to be able to start with friends (Mauss, 1978:183). In another example reported by Yanagita, a merchant from outside the village sells tissue to villagers by saying it was used for the clothing of Kôbô Daishi, a bonze of ninth century who was believed to be still alive by the villagers (Yanagita, 1990:126). The villagers accept the merchant as well as other strangers only when they put them in touch with their gods or ancestors; they don't welcome anyone. Wealth can be acquired only when the order of recollection approves it. The desire to accumulate wealth individually is severely controlled.

The order of recollection functions, thus, as a system of control of ambiguity. Ambiguous existence belongs neither within nor without. It is in the in-between. Ambiguous existence is not socially recognized immediately insofar as the order of recollection strictly controls its entry and exit. The distance between the ambiguous existence and the community can be altered depending on the situation. The ambiguous existence can at times be viewed positively, while at other times the ambiguous existence can be deemed a threat, prompting an unnecessary change within the community.

The situation has gradually changed since the seventeenth century as the capitalist system developed in Western Europe. The capitalist system liberates the desire to appropriate what someone else has produced without going through the order of recollection. Manufacturers and traders are not interested in which person buys their product. Consumers, on their side, don't have to know who has really produced their goods. This desire can be called *capitalist desire*, which keeps people always interested in something new. The quest for the new is openly approved, even recommended. At the same time, accumulation of wealth acquires in itself a positive value, and desire to possess more commodities is liberated. Capitalist desire also transforms the status of objects into that of commodity. According to Karl Marx, an object, when it becomes a commodity, acquires an exchange value besides a use value. An exchange value gives an object an identity as commodity while a use value constitutes its proper difference.

Capitalist system approves, thus, openly, the ambiguity because free trade is an essential condition for the development of the capitalist system. Furthermore, industrial capitalism that develops in the nineteenth century needs manpower whose mobility is assured. The attitude of capitalist system and the order of recollection toward ambiguity is thus completely opposite. Therefore, they contradict each other and cannot coexist. In order for one to survive, the other must crumble.

Capitalist desire and museological desire

For capitalism to develop, it is necessary to deny, or at least to attenuate the force of order of the recollection. The most important question for promoters of capitalism is, therefore, how to make a new social order, which would enable the development of the capitalist system without hindrance.

The first and one of the most important works on this questioning is *Leviathan* (Hobbes, 1651). In his political philosophy, he asks what a human being is. According to Hobbes, human beings are similar. They all think, desire, and have emotions. This similarity always gives rise to a state of war if there is no sovereign power. The state is the sole guarantor of social order, and as such, only the state can bring prosperity.

State-building does not simply mean the creation of political institutions. It requires the creation of a cultural pivot to which the members of the nation can adhere and accord respect. The museum fulfills this function of the cultural pivot in post-revolutionary France. Discourses, which insist on the utility and necessity of the museum, give, at least, four reasons for which museum plays a crucial role in the nation-state. In the first place, museums contribute to commemorate the glory of the revolution and honor the "great persons" in history. Second, museums allow visitors to learn history and art through exhibited objects reserved for few people until then. Third, the state can better preserve objects and buildings that have become cultural heritage. Lastly, the state does not simply transmit objects and historical monuments in a better technical way. In doing so, it puts an end to the conflicts caused by the revolution and inaugurate a new pacific age (Poulot, 1997:117–134).

Museums that display valuables and historic monuments are becoming new centers of society (Franklin, 2019:5–6). And as such, museums allow all members of the nation to share precious objects that make them proud to be its members. This is how they help them to acquire a national identity and become equal beyond each difference. The museum as a cultural pivot of the state helps to create a sense of nationalism. Newly acquired national identity liberates peasants from their community, closed until then. The boundaries of the communities become blurred, and it is easier for the ambiguous existence to enter them. Ambiguous existence circulates freely from one community to another. At the same time, national history becomes important. It is no longer the symbolic exchange between the living and the dead but the transmission of objects and monuments from the past to the future that ensures social order. In the case of the order of recollection, the past is localized in space, whereas in the nation-state, the past is situated in time. Museological desire germinates in this circumstance.

With the development of the museological and capitalist desire, to be socially recognized, an object must acquire the status of a commodity or of a cultural heritage. An object can be a commodity in the capitalist system. It can also acquire cultural value in cultural heritage institutions. For example, an art object can be sold in the art market as a commodity. But when it becomes a part of a museum collection, it is transformed into a cultural heritage. So, in a certain way, these two desires complement each other. Therefore, not only the capitalist system but also cultural heritage institutions liberate the desire to understand the unknown and to appropriate what ambiguous existence has produced. Curators seek an object produced by a society to which they do not belong historically or geographically. Tourists visit a cultural heritage site not familiar to them.

Once an object acquires the status of a commodity or cultural property, it should be exhibited in a public place. Merchandise should be exhibited in shops. In the same way, fine arts should be exhibited in museums and historical monuments be opened to the public. In particular, museological desire highlights what Walter Benjamin called exhibition value (Benjamin, 2005). Benjamin pointed out that in the modern era, exhibition value became more and more valuable, and at the same time, various products are recognized more especially as a work of art. Even religious monuments like churches and cathedrals are first regarded as works of art and as such have to be exhibited for the public. This public nature is assured by the state considering that cultural property belongs to the people. In this sense, museological desire aims at being "democratic".[4] Essentially, the bourgeoisie, the main promoter of the capitalist system, developed museological desire, since the bourgeoisie had the desire for the property of kings and aristocracy on the one hand and that of so-called primitive societies on the other. However, they didn't monopolize objects that they considered valuable. They shared some of them by means of cultural heritage institutions.

Another convergence between the capitalist system and cultural heritage institutions is authenticity and integrity. The notion of authenticity comes from the intention to make a clear distinction between true and false. Cartesian philosophy is the first to emphasize this distinction. According to Descartes, "I had always had an extreme desire to learn to distinguish true from false in order to see clearly into my own actions and to walk with safety in this life" (Descartes, 1966:38). In the capitalist system, diamonds, Levi's jeans, or any merchandise should acquire a sign of legitimacy that they are authentic in one way or the other. It is the same for cultural heritage institutions. An art object should be authentic so that it is judged as having great artistic value. Authenticity is related to the concept of uniqueness. An authentic fine art object constitutes a unique existence and, as such, is regarded as a reference. It is, thus, continually copied and reproduced. The notion of integrity also applies to both. A commodity must be, by definition, integral. If it is defective, a new one should replace it. In the same way, cultural heritage has to remain intact.

Contrary to the elements of convergence, like the desire of possessing what anonymous workers have produced, the need of exhibition, and the value of authenticity, a difference between the two lies in the fact that cultural heritage institutions consist of preserving objects and monuments forever; as soon as an object is classified as cultural heritage, it should be regularly repaired and never destroyed. Cultural heritage is, thus, closer to a mummy, a favorite museum object. As part of conservation efforts, all elements that pose a risk to cultural heritage should be avoided. The environment around cultural heritage sites should, therefore, be clean, transparent, and safe. Any sign of violence must be eliminated. A cultural heritage site thus becomes a symbol of eternity. Another difference resides in the question of value and price. The value of an item is evaluated by its price, while cultural heritage is priceless. When a picture is being sold in the market, it is a commodity. But when it is part of a museum collection, its value cannot be evaluated quantitatively by a price any more. If cultural heritage should be preserved forever, it is because capitalist desire tends to deny the notion of continuity. On the contrary, the capitalist system seeks to produce and sell items as quickly as possible to make more profit. The differences between capitalist desire and museological desire do not, therefore, signify their opposition but their complementarity. Since the capitalist desire tends to deny continuity, the role of cultural heritage institutions is to complement this defect and ensure the continuity of society.

Birth of World Heritage institutions

Museological desire is explicitly drawn from the Convention Concerning the Protection of the World Cultural and Natural Heritage concluded in 1972 (the World Heritage Convention). The convention defines the value that can give rise to museological desire and the conditions in which it can be realized in the Operational Guidelines for the implementation of the World Heritage Convention (UNESCO 2020t). The value is clearly defined as Outstanding Universal Value. It means "cultural and/or natural significance which is so exceptional as to transcend national boundaries and to be of common importance for present and future generations of all humanity". And "as such, the permanent protection of this heritage is of the highest importance to the international community as a whole" (UNESCO, 2013). To be deemed to be of Outstanding Universal Value, a property must satisfy the conditions of integrity and/or authenticity and have an adequate protection and management system to ensure its safety. The Guidelines define strict conditions of authenticity such as form and design; materials and substance; use and function; traditions, techniques, and management systems; location and setting; language and other forms of intangible heritage; spirit and feeling; and other internal and external factors. Authenticity and integrity become the conditions *sine qua non* of preservation.

This idea of worldwide protection of cultural heritage appears explicitly in the Convention for the Protection of Cultural Property in the Event of Armed Conflict, concluded in 1954 in Hague (the 1954 Hague Convention) (UNESCO, 2020a). The convention is directly related to World War II. The convention recognizes the fact that cultural property has suffered grave damage during recent armed conflicts and that, by reason of the developments in the technique of warfare, it is in increasing danger of destruction. By stressing the effects of war on cultural heritage, the convention insists on the need for protection of cultural heritage. Especially, this convention consists in "preventing the exportation from a territory occupied".

The Hague Convention deals with damage to cultural heritage caused by war. But the relationship between cultural heritage institutions and war is deeper and goes back to the time when cultural heritage institutions were just born. The conflict involving valuable objects appears precisely just after the revolution when Napoleon Bonaparte led the Egyptian expedition (1798–1801). From a geopolitical point of view, the objective of the French army was to conquer Egypt in order to block the route from India to Great Britain. The expedition therefore inevitably provoked fights between France and Great Britain. The fight was not limited to the territorial acquisition, but it took place for the acquisition of the precious objects. Bonaparte took with him to Egypt a "science and arts commission", composed of some 160 civilian technicians, engineers, and scholars. The most famous discovery of the commission was the Rosetta Stone. If Bonaparte and his commission were interested in the Ancient Egyptian civilization, this was because the Ancient Egyptian civilization was considered a cultural origin of humanity. Egypt was not radically different from France. French civilization should have, in a sense, a root in the Ancient Egyptian civilization. British people shared this conception of civilization. The result was the battle to obtain the Rosetta Stone. Great Britain defeated France and brought the Rosetta Stone to London in 1801. The Stone is on display at the British Museum since 1802.[5]

While two nation-states come into conflict over a territory, there is also a conflict over cultural products that do not belong to these two nations. The acquisition of objects by museums is, thus, not without violence. On the contrary, it goes hand in hand with the pursuit of the nation-state to expand its territory. And it is not by chance. The logical link between the two is clearly exposed by Thomas Stanford Raffles, colonizer of Singapore, and described by Anderson as "ominous emissary". His idea of colonization was not a simple policy of domination. Influenced by the Enlightenment, he thought that the construction of a colony was to become a zone for free trade, and the development of the trade should go hand in hand with that of the education. Colonizers should study and comprehend deeply native inhabitants, who should, in their turn, learn and understand their own culture at school; when he visited Malacca in 1811, he had the idea of creating Malay's schools in the future. The originality of a colonization policy like that of Raffles lies in the fact that the colonization is not a simple conquest by military forces, but the colonizers have to know and comprehend the culture of the people they plan to colonize. Raffles wrote a book on the Java story (Raffles, 1988). Furthermore, in the plan of colonization, colonized people have to learn their culture on their side. This "culture", constituted by colonizers, called *Orientalism* by Edward Said (Said, 2003), would be shared by a colonized people through the education system.

The conflict over the Rosetta Stone and Ruffles' civilizing ideology shows another dimension of cultural heritage institutions. They go hand in hand with the expansionist tendency of the nation-state. We have already seen that they served as the pivot of the nation-state and became a matrix of nationalism. Their first purpose was to establish social order. But this nationalism was not only aimed at the interior of the nation. From the outset, it included the expansionist tendency.

Why then does the nation-state accompany the expansionist tendency? To answer this question, first of all, the relationship between society and space should be take into account. Society needs borders to circumscribe this space of domination. The order of recollection put the first borders between the living and the dead by fixing the space in which the spirits of the dead or the gods exist. Nevertheless, the symbolic exchange between the two takes place regularly in religious premises. Death does not mean final separation. At the same time, the second borders are installed between the inside and the outside of the society. The interior is a space directly led to the world of the spirits of the dead or to the gods, while the exterior is a relegated world; beyond the second borders, there are huge areas that members of a society don't really know or are indifferent to. The relationships between community space and the presumed world where the dead are present are the most important, and the space beyond the second borders hardly attracts the attention of the community.

On the contrary, the nation-state installs the first borders between the inside and the outside of the nation in strictly circumscribing their borders with neighboring countries. Those who are outside the first borders are recognized as "foreigners". The first borders are, in a way, established not vertically, like the order of the recollection, but horizontally, and they are represented geographically. At the same time, the second borders are implicitly installed between the outside, represented by the first borders, and space which is not yet represented. Those who are outside the second borders are not completely different. They are like us because they are human, and as such they can be familiar to us. This is how, contrary to the order of recollection, the nation-state takes an active interest in beyond the second borders and seeks to know more about space beyond the second borders. Otherness is represented geographically.

But this interest in otherness is not neutral and disinterested. On the contrary, the nation-state is always ready to grow. The nation-state intends to annex what does not belong to it. Boundaries are not definitively fixed. Each nation-state strives to enlarge its territory. It inevitably follows that war arises between nation-states. Why then is the nation-state constantly seeking to grow? To answer this question, we will refer to Thomas Hobbes. According to him, in the state of nature, all human beings are equal in their abilities, which inevitably leads to struggle. In order to overcome this warlike state, only by forming a state can human beings secure security and private property.

At first glance, the argument seems to have a certain consistency, but there is a fundamental contradiction lurking in it. In the first part, it is shown that human is a universal concept and that there is no significant difference among individual humans. Needless to say, it is assumed here that human beings do not exist only in one particular area but live everywhere in the world. The notion of human in the state of nature advanced by Hobbes must apply to all human beings without exception. However, in the second part, when individuals transfer to the state the complete freedom they possessed in the state of nature, the discussion is focused only on the establishment of a single state. The fact that there are other human beings outside of the state, who also form the state, has somehow disappeared. The state is founded on the premise of human identity. But only certain people have national identity, formal membership in a state, under the universality of human identity. Somehow, the universal human identity has been transformed into a more limited national identity. This is not so much a leap in Hobbes' logic, as it is a fundamental contradiction in the modern state itself. The inclusive tendency of the nation-state is born in order to overcome the contradiction between the human identity and national identity and to reduce the difference between national space and space which does not belong to the nation, as if all human beings should ultimately belong to the same nation. If all nations seek to expand their territory and dominate the entire world, the inevitable result is a state of war. This is what really happened. A series of wars took place until the two world wars.

The museological desire develops within the framework of this expansionist tendency of the nation. The nation-state, the guardian of human civilization, has the right to appropriate objects produced elsewhere if it has discovered them. The national museum becomes a place par excellence which collects and displays cultural works from around the world. Archeology, history, and anthropology become important disciplines for the choice of the collection and the staging of the exhibition. The culture of other societies is therefore our culture because it belongs to so-called culture of humanity. On the other hand, the museum constitutes, in a way, a space of compromise, which illusively realizes a united world by bringing together cultural objects from around the world, a world that a nation-state could not create in reality.

That said, a national museum can never become a museum for the whole of humanity. UNESCO therefore seeks to create authentically universal institutions for the protection of natural and cultural heritage. With the conclusion of the World Heritage Convention, it is for the first time that cultural heritage institutions external to all nation-states have been founded. The World Heritage Commission represents all of humanity. It is no longer such and such a nation, but it is the international organization which has the right to give value to cultural heritage in the last instance. The creation of World Heritage institutions transforms, thus, the relationship between cultural heritage and society insofar as international institutions establish criteria for cultural heritage. The conclusion of the Convention is a turning point for the treatment of cultural heritage.

War, space, cultural heritage

The World Heritage Convention qualitatively transforms the nature of heritage by introducing the notion of territory. The 1954 Hague Convention already introduces the notion of territory. It considers that a cultural property does not exist independently from a territory. It always belongs to a place where it was produced. The notion of territory links cultural heritage and the environment in which it was produced. The World Heritage Convention concluded in 1972 succeeds to the spirit of the 1954 Hague Convention. In the whole articles of the World Heritage Convention, the word territory is found 11 times. Cultural heritage can only exist in relation to a defined territory.

The World Heritage Convention brings more radical change in the matter. In fact, it gives more precision to the relation of cultural works to space by introducing the concept of cultural heritage. The word property is reminiscent of cultural works that belong to an individual or a group. It cannot take sufficient account of the relationship between cultural products and the environment in which it is located. The word heritage is more abstract and no longer signifies a specific object. In the World Heritage Convention, cultural heritage is classified into three categories in Article 1. The first category is monuments. These are "architectural works, works of monumental sculpture and painting, elements or structures of an archaeological nature, inscriptions, cave dwellings and combinations of features, which are of outstanding universal value from the point of view of history, art or science". The second category is groups of buildings, "groups of separate or connected buildings which, because of their architecture, their homogeneity or their place in the landscape, are of outstanding universal value from the point of view of history, art or science". The third category is sites. Sites mean "works of man or the combined works of nature and man, and areas including archaeological sites which are of outstanding universal value from the historical, aesthetic, ethnological or anthropological point of view" (UNESCO, 2020b). All three types are related to space. In particular, in the category of groups of buildings, there are some groups of buildings that have an outstanding value, because of "their place in the landscape". In this type, the relationship between groups of buildings and the landscape gives an outstanding value. The notion of sites also shows the explicit relationship between cultural heritage and space insofar as sites include "areas".

Furthermore, every world heritage site should have a core zone and a buffer zone. Inscription of a cultural heritage on the World Heritage List thus signifies literally a production of space. The area where a world heritage site is situated should be planned so that it will be well preserved. To do this, not only core zone in which the world heritage site is situated but also the buffer zone is created. According to Francesco Bandarin,

> buffer zones are an important tool for conservation of properties inscribed on the World Heritage List. All along the history of implementation of the World Heritage Convention, the protection of the "surroundings" of the inscribed properties was considered an essential component of the conservation strategy, for cultural and natural sites alike.
>
> (Bandarin, 2008:9)

Zoning is adopted as a strategy for the protection of world heritage sites. According to the terminology of Henri Lefebvre, zoning is a sort of *representation of space*[6] which contributes to the production of space. Practitioners of world heritage plan and carry out zoning like urban planners.

When museological desire appears, it is rather related to time. Valuable objects have their history, and this history illustrates the history of the nation. But especially with the World Heritage Convention, museological desire goes not only to the objects but also to the environment in which they are located. This emphasis on the spatial aspect of cultural heritage seems to be close to the military conception of space. In particular, the notion of zoning means the defense of a delimited territory as the military defense of a territory. The analogy of this strategy of heritage protection to military tactics is not accidental. World Heritage institutions have to deal with the elements that threaten World Heritage sites and must therefore function even in times of armed conflict. World Heritage institutions that emerged from World War II are linked to war on several levels.

Notes

1 Before Mauss, Friedrich Nietzche made a similar observation in *Genealogy of Morality*:

> the living generation always acknowledged a legal obligation towards the earlier generation, and in particular towards the earliest, which founded the tribe (abbrev) . . . There is a prevailing conviction that the tribe *exists* only because of the sacrifices and deeds of the forefathers, – and that these have to be *paid back* with sacrifices and deeds.
>
> (Nietzche, 1994:65)

2 Les pèlerins de l'eau vive (The pilgrims of the living water) is an association which consists in curing alcoholics by faith in Jesus.
3 Tono is a market town in the north of Japan where the Japanese ethnologist Kunio Yanagita made his investigation.

4 It is also interesting to note that exhibition value develops with the technique of reproduction. The capitalist system advances this technique.

5 The case like what happened during the Egyptian expedition by Bonaparte, that is, exportation of Rosetta Stone, is prohibited by the 1954 Hague Convention. And yet, the Rosetta Stone is on display in the British Museum.

6 Representation of space means planned space. It constitutes dominant space in society (Lefebvre, 2000:48).

2 War and world heritage

The expansionism of nation-states culminated in the world wars, whose main target was no longer humans but the environment (Sloterdijk, 2009:14). Attack on the environment by the air forces destroys towns and cities and creates considerable victims, including civilians. It even happens that the whole city is demolished. As cities become battlegrounds and targets of attack, inhabitants no longer feel secure. The bombardment is not simply physical destruction, but it is a profound attack on the identity of the space. And the feeling of insecurity and fear may remain after the end of the war. This characteristic of war since the twentieth century involves three fundamental questions. (1) How can we avoid great wars? (2) How are nations and cities recovering after the experience of bombardment? (3) How can the feeling of insecurity be dispelled, and the trauma caused by war can be taken care of?

These questions are all related to cultural heritage. With regard to the first question, after World War II, the system of conflict resolution has been established by the foundation of United Nations. The tension between the United States and the USSR certainly continued, but there were at least efforts to avoid the outbreak of the third world war. UNESCO, United Nations Educational, Scientific and Cultural Organization, is a part of United Nations, and as such, its vocation is the realization of peace in the world. The World Heritage Convention was concluded within the framework of UNESCO's objective. The World Heritage Convention declared then, at the beginning of its text, that the cultural heritage and the natural heritage are increasingly threatened with destruction not only by the traditional causes of decay but also by changing social and economic conditions which aggravate the situation with even more formidable phenomena of damage or destruction (UNESCO, 2020b). Among the causes that endanger World Heritage sites, "the outbreak or the threat of an armed conflict" is mentioned. If a region that has a World Heritage site experiences armed conflict, the site will be placed on the List of World Heritage in Danger. One of the aims of World Heritage institutions is to protect World Heritage sites threatened by war.

World Heritage institutions are more directly related to the second question concerning the reconstruction of the society. The destroyed city visualizes uncertainty, physically and explicitly. It should not be left as is, because if it is, the future of the area is uncertain. In many cases, reconstruction does not advance quickly.

DOI: 10.4324/9781003014904-3

But if nothing is done, this kind of space produces a disorder in the area. To restore the social order, the area should be spatially reconstructed. In this sense, war not only causes destruction but also gives rise to the production of the space.

There are two types of World Heritage sites concerning the reconstruction of post-war society. The first type is the case in which the inscription on the World Heritage List goes hand in hand with the reconstruction of the society. In this type, one of the purposes of inscription is peacemaking. The first type is, therefore, directly related to the first question: how can we avoid great wars? A subtype of this first type is the site that was once on the List of World Heritage in Danger and then was removed from it. The second type is the case in which the process of reconstruction itself becomes recognized and inscribed on the World Heritage List. As we will see, this type goes so far as to transform the traditional criteria of cultural heritage that do not give value to the reconstruction of monuments; they must keep the original state. We will take as examples of the first type the Angkor in Cambodia, the subtype Dubrovnik in Croatia, and the second type Warsaw in Poland.

The third question relates to the second question. The examples of the reconstruction of post-war society inevitably raise the question of the negative memory of war. A new idea then emerges on the treatment of monuments that recall negative memory. This idea suggests preserving these monuments instead of demolishing them. The bombed monuments can then be transformed into historical and heritage monuments. These monuments are called in Japan *hu no isan*, which means negative heritage. They include monuments or memorials of war, buildings from colonial times and abandoned mines that have caused environmental pollution or industrial accidents. As examples of negative heritage, we will take two cases concerning nuclear weapons, the Genbaku Dome and the Bikini Atoll.

Reconstruction of Angkor Wat and Dubrovnik

Angkor is a typical example in which the registration for the World Heritage List goes hand in hand with the reconstruction of society.

At the beginning of the twentieth century, Angkor was a forgotten ruin. We know from a novel of André Malreaux, *Voie Royale* (Malreaux, 1930), that Western vandals were seeking the unexplored heritage of the kingdom. The heroes of the novel finally arrive at Angkor Wat (Figure 2.1), almost inapproachable at that time, and try to take parts of the gigantic vestige. But during this adventure, one of them is injured, and they are forced to go through "free" areas, where they felt threatened by the presence of the Moi and Steing people, who had total autonomy. Before independence, when Cambodia was under the French protectorate, there were areas that were not subject to the protectorate.

When Cambodia won its independence, it was difficult to establish a stable state order, and this situation gave rise to the birth of the Khmer Rouge. The Khmer Rouge almost destroyed temples that they considered traces of a bad tradition. But curiously, Angkor Wat escaped demolition as the Khmer Rouge did not attach much importance to it. We know in retrospect that the Khmer Rouge

Figure 2.1 Tourists climb to the summit of Angkor Wat
Source: Photo by the author.

perpetrated an autogenocide during its reign. A large number of Cambodian people were killed by their compatriots between 1975 and 1979. In 1979, the Vietnamese army overthrew the Khmer Rouge. But the group ran into the jungle and continued to resist the new government supported by Vietnam. Before and during its escape, the Khmer Rouge laid a great number of mines that would create many victims.

After more than ten years of Civil War, the 1991 Paris Peace Accords were signed. One year later, Angkor was registered for the World Heritage List before the establishment of the new government in 1993. Therefore, the registration for the World Heritage List coincides with the reconstruction of Cambodia after the Civil War. We could even say that registration aimed to promote the reconstruction of Cambodian society destroyed by the Civil War.

The Authority for the Protection of the Site and the Management of the Angkor Region (APSARA), founded in 1995, tried not only to restore the heritage but also to improve the economic situation of the region in collaboration with foreign countries and international organizations. Despite the registration for the World Heritage List, water and wood are lacking in the area, and local people remain poor. Local people illegally cut down trees for charcoal. APSARA took measures to alleviate the lack of water and wood and to provide electricity. It set up solar collectors and distributed batteries to each family. APSARA also promoted "community tourism". For example, Banteay Srey community tourism (Figure 2.2) was established in 2012. Members of the community manage tours for boat

Figure 2.2 The Banteay Srey Community Tourism
Source: Photo by the author.

rides and fishing in collaboration with APSARA and New Zealand. "The Banteay Srey Community Tourism initiative is set up by the community and for the community. Money raised from the tour goes to the local community[1]". Moreover, APSARA played the role of educator. It taught elementary, junior, and high school students the importance of preserving Angkor. It also asked monks to collaborate, because Cambodian people deeply respect monks as religious and moral authorities. If monks attribute significance to the World Heritage designation and support its preservation, local people faithfully obey their instructions. ASPARA thus leads activities for autogenous development that aim to realize development for local people by local people.

However, the tourist industry, which seeks profits, expands independently of the politics of autogenous development. The number of foreign tourists increased greatly (from 118,183 to 6,610,592) in 26 years (from 1993 to 2019) (National Institute of Tourism, Tourism Statistics Department, 2020). The economy of Cambodia, thus, largely depends on tourism today. In Siem Reap, a town close to Angkor, many modern hotels are built for foreign tourists. An area named Pub Street gives the impression of being in London because there are only foreign visitors essentially from Europe and United States.

An elementary school is situated next door to the World Heritage site. Some children go to sell souvenirs just after school. Cambodian people who work in tourism and especially want to work are obliged to learn English. After the end of French colonization, the dictatorship of the Khmer Rouge, and the Civil War, they are obliged to learn the language of a country that bombed Cambodia,[2] as if Cambodian people should always assimilate the culture of others. Anyway, the risk that visitors run is minimized, and the inhabitants of the city are well disciplined to accommodate foreign tourists. Many inhabitants live inside the sites and have opportunities to meet visitors in their daily life. Tourism reduces, therefore, the distance between local people and foreigners.

This reduction in distance leads the inhabitants to share the external point of view. They begin to look at their world, to some extent, as strangers. This is what we have called the doubling of the world. At the same time, it reduces the autonomy of local life. World Heritage institutions impose strict rules. People who reside in the preservation district must conform to the regulations of the law on the protection of the natural and cultural heritage promulgated on January 25, 1996. They are not even at liberty to reform their dwellings or to construct a new one. Some people are suggested to move out to a new village constructed out of the buffer zone of the site. To preserve the site perfectly, the life of local residents is more or less sacrificed. Furthermore, the economic leap, thanks to the development of tourism, has not managed to remove the negative traces of the Civil War. Inside the World Heritage site of Angkor, victims of mines laid by the Khmer Rouge play traditional music. In Siem Reap, there is the Killing Fields (Figure 2.3), a memorial tower for the victims of genocide. Inside the tower, we can see many skulls. The development of the city cannot completely erase the remnants of genocide. The driver who drove us to the Killing Fields told us about his experiences during the Khmer Rouge era.

Figure 2.3 Killing fields
Source: Photo by the author.

Figure 2.4 Tourists around the Great Onofrio Fountain in Dubrovnik
Source: Photo by the author.

He had been in a labor camp where there was little food. The workers ate frogs in secret, but if supervisors discovered who ate the frogs, they shot him. He had a narrow escape and joined the police force trained by the Vietnamese army.

The Old City of Dubrovnik constitutes a subtype of the first type; the site was once on the List of World Heritage in Danger and then was removed from it. Dubrovnik experienced a great earthquake in 1967. Despite this event, the city "managed to preserve its beautiful Gothic, Renaissance and Baroque churches, monasteries, palaces and fountains"(UNESCO, 2020c) (Figure 2.4). It is for this reason that The Old City of Dubrovnik was inscribed on the World Heritage List in 1979. But the Croatian War of Independence took place from 1991 to 1995. Due to this armed conflict, in 1991, the Croatian site was inscribed on the List of the World Heritage in Danger. In 1998, the site was considerably restored, and it was removed from this list. When we visited in 2013, the Old City attracted a lot of tourists, and the traces of the war were not visible.

The old town of Warsaw

The Old Town of Warsaw is an illustrated example in which the process of reconstruction itself becomes recognized and inscribed on the World Heritage List.

The Old Town of Warsaw was almost destroyed by Nazi troops in August 1944. The town, which had been destroyed by hard battles, was reconstructed in order to revive the city of the past. The Market Square and St. John's Basilica of the Old Town were successively restored in 1966 though the restoration of the Royal Castle was delayed because of the hesitation of the communist regime. It was only in 1971 that the restoration of the Castle started thanks to the financial contribution of Polish people and Poles abroad (Wolska, 2013).

Once the reconstruction was completed, the Polish Government submitted the nomination file for the registration of Warsaw at UNESCO. However, the inscription of the Historic Centre of Warsaw on the World Heritage List did not proceed without opposition. During the first attempt to secure the inscription in 1978, International Council on Munuments and Sites (ICOMOS) noted the Warsaw proposal needed further expert study. In the next year, the Bureau of the World Heritage Committee recommended deferral (Cameron, 2008:19). At the second attempt, ICOMOS was placed in a favorable position for the inscription of Polish site, and the World Heritage Committee recognized it as "an outstanding example of a near-total reconstruction of a span of history covering the 13th to the 20th century". The Historic Centre of Warsaw was, thus, registered for the World Heritage List while in the 1980 version of the Operational Guidelines, the criterion of authenticity was revised to include the sentence: "reconstruction is only acceptable if it is carried out on the basis of complete and detailed documentation on the original and to no extent on conjecture" (Labadi and Colin, 2010:68).

According to the Operational Guidelines for the Implementation of the World Heritage Convention, World Heritage sites need to satisfy the following conditions of authenticity: form and design; materials and substance; use and function; traditions, techniques, and management systems; location and setting; language and other forms of intangible heritage; spirit and feeling; and other internal and external factors. To faithfully respect the Guidelines, every effort must be made to maintain the "authentic" state of the World Heritage site. But in fact, it seems extremely difficult to fulfill all these conditions, insofar as all objects and architecture deteriorate, and restorationists cannot always obtain original materials. If the conditions of authenticity outlined by UNESCO needed to be faithfully respected, there would not be many truly authentic monuments or objects. In reality, restorationists use synthetic materials if they cannot find original materials. In this regard, the inscription of Warsaw's Historic Centre in 1980 was a turning point. It established the principle that World Heritage institutions should not only accept but also appreciate reconstructions. The reproduction acquired the same status of authenticity as the original if the required conditions are satisfied. This is the strategy of reproduction which makes it possible to restore threatened cultural heritage and in certain cases the reproduction will substitute the original.

Taking account of this new mode of preservation and conception of authenticity, the Nara meeting was held in 1994 on the initiative of the Japanese Government. The result of the meeting was stated in the Nara Document of 1994. The Nara Document on Authenticity states in an abstract way that all value judgments attributed to cultural properties may differ from culture to culture and even within the same culture. Thus, it is not possible to base judgments of values and authenticity on fixed criteria. The document adds that on the contrary, the respect due to all cultures requires that heritage properties must be considered and judged within the cultural contexts to which they belong. This idea about the diversity of cultural values stated in the Nara Document was finally enacted in the revised Guidelines of 2005. In the same year, the Old Bridge Area of Mostar that had been damaged by the Bosnian War was registered for the World Heritage List, because the reconstruction of the area was appreciated in the same way as the Historic Centre of Warsaw (Cameron, 2008:22).

The model of reconstruction in the Warsaw case resulted from the bombing of the area. The reconstruction of a society that suffered from war thus modified the modality of preservation and legitimated a different type of cultural value based on the strategy of reproduction.

Negative heritage

The second part of this chapter will discuss the issue of negative heritage. As noted at the beginning of this chapter, in both world wars the attack is on the environment. For example, the atomic bombing targets and destroys the environment of the entire city. Peter Sloterdijk calls the bombings in Hiroshima and Nagasaki "radioterrorism". The A-bomb causes immediately a large number of victims of heat. This is the first effect of radioterrorism. But radioterrorism continues to make another type of victims, victims of radioactivity. Even if the reconstruction of the destroyed city is almost complete, the recovery of the society is not finished. The A-bomb causes effects for a very long time, and many victims suffer from disease like leukemia due to radioactive contamination. The attack on the environment produces damage over a long period of time. In this sense, it is much more effective. Furthermore, the memory of the bombing does not disappear easily among the victims. Victims prefer to quickly forget memories that invoke abhorrent feelings. Even if they are unable to fully forget these matters, they want to distance themselves from them as much as possible. Monuments that bring back bad memories should disappear as soon as possible.

However, a new principle is emerging, that of preserving monuments affected by the bombing as war heritage. This is one of the characteristics of the Age of Preservation, in which everything can become cultural heritage. In such a way, instead of forgetting them, museological desire suggests constantly and faithfully recording the negative experiences of those who lost their loved ones. One of the first illustrations of the war heritage that preserves tragedy is Oradour sur Glane in France. It is a village in southwestern France where most of the inhabitants were killed by a Waffen SS unit. During his visit in March 1945, General de Gaulle recalled "that a place like this remains something common to all, something where everyone recognizes the common misfortune, the common will and the common hope". In 1946, the ruins of Oradour were classified as a historical monument.

In the same year, the Polish Government decided to preserve Auschwitz-Birkenau. The Museum opened the next year, and it was inscribed on the World Heritage List in 1979. Listing had been started just the year before. The site was registered based on Criterion VI of UNESCO. Criterion VI is to be directly or tangibly associated with events or living traditions, with ideas, or with beliefs, with artistic and literary works of outstanding universal significance. Auschwitz-Birkenau was constructed as a concentration camp and used for extermination. It shows "irrefutable evidence to one of the greatest crimes ever perpetrated against humanity" (UNESCO, 2020d).

Another negative war heritage inscribed on the World Heritage List is the Genbaku Dome (Figure 2.5), aka Hiroshima Peace Memorial, in Hiroshima. It was registered for the World Heritage List in 1996 only based on criterion VI. The Genbaku Dome was the Hiroshima Prefectural Industrial Promotional Hall and not a military facility. It was situated just beside the hypocenter of the blast on August 6, 1945. The Outstanding Universal Value of the Genbaku Dome is considered "a stark and powerful symbol of the achievement of world peace" (UNESCO, 2020e).

The atomic bombing took place on August 6, 1945 in Hiroshima and on August 9 in Nagasaki. Japan surrendered unconditionally six days later. And only on September 2 of the same year, the prefecture of Hiroshima decided to preserve the area around the hypocenter as a memorial area. An Australian Lieutenant

Figure 2.5 Genbaku Dome
Source: Photo by the author.

Commander suggested choosing 13 Genbaku monuments that should be tourist attractions, including the Genbaku Dome. Japan was occupied by the Allies, and the idea for the conservation of the dome would have come from the occupiers. In fact, as mentioned in the Introduction, before World War II, museological desire was almost absent in Japan. The idea of war heritage really did not exist among the Japanese. On the contrary, for the Americans, atomic bombing put an end to the war. The dome could therefore become the symbol of peace. It was worth preserving. The victims of the atomic bomb would not share this idea, but defeated Japan could not openly criticize the atomic bombings as a great crime. The criticism at the time was mainly directed at the Japanese army and the government led by the army officials.

Hiroshima city, which should reconstruct the destructed city, adopted the suggestion of the Australian Lieutenant. Then, still under the occupation of the Allies, in 1949, the project of the pacific memorial park designed by Kenzô Tange was adopted in which the museum and the memorial (Figure 2.6) would be built in such a way as to connect them to the Genbaku Dome on the same line. The dome was only one building that remained in this area. However, the Genbaku Dome was not immediately designated as a monument to be preserved. Hiroshima city could not immediately make the final decision on the fate of the dome. Unlike

Figure 2.6 Memorial Cenotaph for the A-Bomb Victims and in the background, Genbaku Dome

Source: Photo by the author.

Auschwitz-Birkenau, the Genbaku Dome was on probation for a long time. This is essentially due to the ambivalent nature of the dome. It was indeed difficult to give a unique and legitimate meaning to the Genbaku Dome for two reasons. First, there were survivors of the atomic bombing who did not want to see the striking trace of the bombing. There were those who thought it should be quickly removed precisely because it invoked nightmarish memories.

Second, there were diverse opinions about who was responsible for the bombing. To whom can it be attributed? Should it be attributed to the Japanese army that started the war against the Americans? Or should it be blamed on the excesses of the US army and the US Government? Anger toward the atomic bombing, even if it existed, could not be openly declared, at least not until 1952, when Japan regained independence. The answers to all these questions are diverse and they change over time. Different opinions on the conservation of the Genbaku Dome and the memory of the Pacific War sometimes provoke controversy. The case of the Genbaku Dome shows how a monument that was considered extremely negative has come to be definitely conserved. We will show this process in two parts. In the first step, a historical overview up to the inscription of the dome on the World Heritage List will be described. In the second step, the conflict that arose around the Genbaku Dome and other war monuments will be treated in terms of representation.

The Lucky Dragon and *Orizuru No Kai*

While the unstable and antagonistic situation continued in Hiroshima, another serious incident in the history of nuclear contamination occurred in 1954. The Lucky Dragon, a fishing boat from Yaizu in Shizuoka prefecture, was caught in the atomic tests held at Bikini Atoll. The crew aboard the Lucky Dragon, Daigo Fukuryu Maru, experienced a similar situation as the A-bomb survivors of Hiroshima and Nagasaki, *hibakusha*. They had to keep silent to avoid social discrimination. Silence is a particularly good option if the number of victims is limited. The crew said that immediately after the incident, they were "treated like criminals".

> Fish shops and sushi restaurants said this because the tuna was contaminated, meaning they couldn't sell it. When we passed by, people said to us that it was our fault they couldn't sell anything, and that we'd put them out of business.
> (Statement by Matahachi Oishi, former Lucky Dragon crewman)
> (Lucky Dragon Peace Association, 1986)

Furthermore,

> when we met other crewmen, we hardly ever talked about the incident, and none of us ventured the information that we were on the Lucky Dragon. This was very hurtful, and no one wants to keep opening old wounds, do they?
> (Statement by Fishing Chief Yoshio Misaki, former Lucky Dragon crewman)
> (Lucky Dragon Peace Association, 1986:78)

Given circumstances such as these, it is unimaginable that the affected crewmen would voluntarily talk about their experiences or start a campaign to preserve the Lucky Dragon. The memory of that radioactive fallout is one that they want to get as far away from as possible. Matahachi Oishi, for example, left his hometown of Shizuoka and moved to Tokyo, where nobody knew him.

Contrary to the crew's experience, the Lucky Dragon incident gave rise to anti-nuclear movements. Various support groups for *hibakusha* were established. Support groups and mass media encouraged *hibakusha* to talk about their experiences to transmit them to the future generations. Following this idea, the Hiroshima Peace Memorial Museum was founded in 1955. It was not the same for the Genbaku Dome. The controversy continued, and Hiroshima city did not take the final decision on the preservation of the dome. A change occurred when a support group for the child victims of the atomic bomb, *Orizuru No Kai* (paper crane club), expressed support for preservation in 1960. *Orizuru No Kai* was born from the movement for building a statue of Sadako Sasaki (Figure 2.7), an A-bomb survivor. Ten years after the bombing, she developed leukemia. She was then 12 years old. She began making 1,000 origami cranes to pray for her recovery; there is a custom of making them when Japanese pray for a wish to come true. Unfortunately, Sadako passed away one year later. To mourn her death,

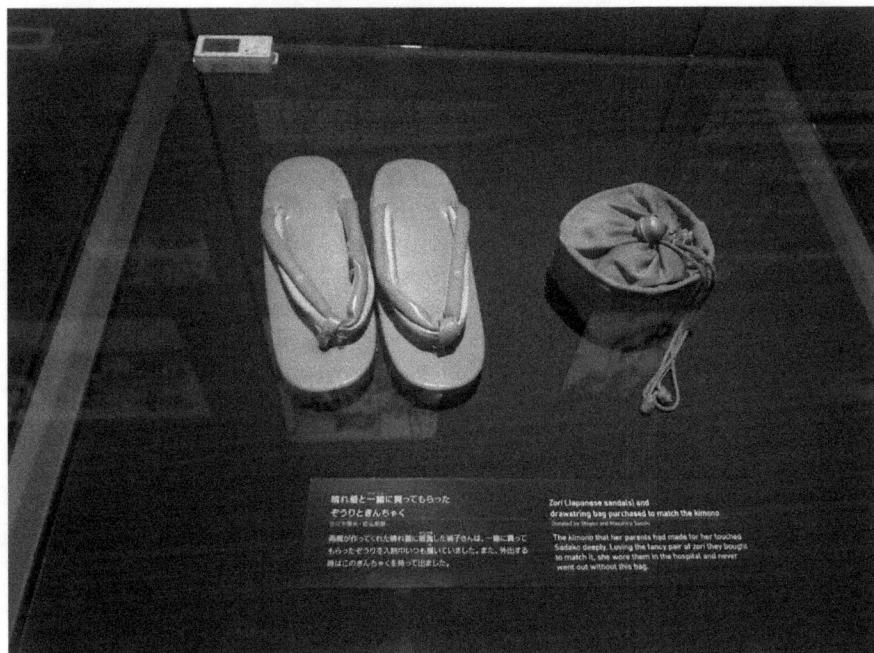

Figure 2.7 Japanese sandal and bag used by Sadako during her hospitalization exhibited at the Hiroshima Peace Museum. Donation by Shigeo Sasaki and Masahiro Sasaki

Source: Photo by the author.

her classmates at elementary school had the idea to erect a statue of Sadako and began a fundraising campaign. In 1958, her statue was built at the Hiroshima Peace Memorial Park. At the same time, *Orizuru No Kai* was founded, and its opinions were influential.

Around 1960, Hiroshima city saw a turning point. The city was more or less rebuilt, and the traces of the bombardment would disappear. Only Genbaku Dome and a few others recalled the tragedy. There was thus a marked imbalance between the two landscapes. This imbalance prevented the re-establishment of the identity of space in Hiroshima. The Genbaku Dome indeed symbolized uncertainty. The dome remained as it had been just after the bombing. The people of Hiroshima did not know what the fate of the dome would be. What then are the strategies that make it possible to dispel the uncertainty embodied by the Genbaku Dome?

It was *Orizuru No Kai* that first expressed support for the preservation. *Orizuru No Kai* discovered, in a diary of Hiroko Kajiyama, who had died at the age of 16 due to leukemia, a passage concerning preservation of the Genbaku Dome. She wrote that only the preservation of Memorial Cenotaph for the A-Bomb Victims and the Genbaku Dome will transmit the horror and the terror of the A-bomb. In Hiroko's idea of always recalling the experience of the world becoming hell, there was an attempt to make sense of the almost ruined dome then. Moreover, various antinuclear associations agreed on the final preservation of the dome in 1964.

Hiroshima city did not immediately make a decision for a practical reason. A monument inevitably deteriorates, and it is necessary to restore it eventually. This restoration requires major financing. The mayor of Hiroshima refused to finance the cost of restoration with the budget of the city government and decided to call on the donation of citizens in 1967. Fortunately, citizen contributions reached the sum necessary for the restoration. In 1968, the mayor, Setsuo Yamada, then drew up a project to make the Peace Memorial Park a "sacred place" to console the souls of the victims by removing all the elements that surrounded the Genbaku Dome, such as the parking lot and shops. He had recourse, in a way, to the order of recollection. Indeed, the main objective was mourning for the victims. And the families of the victims are the main stakeholders. But a little later in 1981, the Genbaku Dome was registered by Hiroshima city for a historic site suffered by atomic bombardment. This means that for the first time, the Genbaku Dome has been considered as cultural heritage. This historical and cultural aspect of the dome was reinforced by the anti-nuclear movements, which made the dome the symbol of peace. So, there are two different meanings given to the dome. The first focuses on mourning for the victims. This position can be qualified as the particularist tendency. The second emphasizes the wish for world peace. This is the universalist tendency (Ejima, 1977). The universalist tendency seeks to share the A-bomb experiences with those who are not A-bomb survivors and families of victims. This tendency was already explicit during the fundraising campaign organized in 1967 for the preservation of the dome. Hiroshima city has raised the necessary funds for preservation with the help of citizens across Japan.

Toward the inscription of Genbaku Dome on the World Heritage List

The universalist tendency has gradually become more pronounced. And the goal was set to have the Genbaku Dome inscribed on the World Heritage List around 1988, when a member of Parliament raised the question of the possibility of concluding the World Heritage Convention using the Genbaku Dome as a possible candidate for the World Heritage site. In fact, just after the Japanese Government ratified the World Heritage Convention in 1992, Hiroshima city decided to make the dome a World Heritage site. But the Agency for Cultural Affairs was not in favor of the Genbaku Dome's candidacy for registration on the cultural property list as the Law for the Protection of Cultural Property does not cover modern buildings. At that time, the Agency for Cultural Affairs considered that the designation of historic sites, *shiseki*, was limited to the monuments built until the middle of 1880s. But the dome was constructed in 1915.

To overcome this situation, the Hiroshima branch of *Rengo*, Japanese Trade Union Confederation, started the movement to make Genbaku Dome a World Heritage site and tried to organize an association to do this. The Hiroshima branch called on several different organizations to participate in the movement regardless of political tendency. This is how the association of lawyers and that of doctors, etc., took part in the movement, and in 1993 the Association for the Promotion of the Genbaku Dome for the World Heritage site was founded. The mayor of Hiroshima and the chairman of the city council were invited to the association's general meeting. Hiroshima city as well as Hiroshima prefecture has therefore agreed almost unanimously on this movement.

The association actively conducted the campaign to collect signatures for the petition. Fifteen deputies from different parties elected in the constituencies of Hiroshima put to the parliament the petition for the project of inscription on the World Heritage List with more than 1,050,000 signatories. In the House of Councilors, the petition was unanimously adopted. But, the Education Committee of the House of Representatives reserved to vote, because some members gave the opinion that value of the Genbaku Dome was not definitely fixed, and it was a little early for registration on the World Heritage List. Not at all discouraged, the association continued lobbying the deputies concerned, and finally Prime Minister Tsutomu Hata then decided to accept the association's request. The House of Representatives thus adopted the petition in June 1993. For it to be a historic site, the agency has changed its rules, and the target date for designation as a historic site was around the end of World War II. The World Heritage listing project transformed the design of the historic site. And thanks to all this series of efforts, in 1996, the dome was inscribed on the World Heritage List despite opposition from the United States and China (Jidai eno messêji, 1997:42–155).

Until the inscription, Hiroshima city, owner of the Genbaku Dome, was primarily responsible for its preservation. But after that, the Ministry of Health, Labour, and Welfare started to collect testimonies from survivors of the Hiroshima and Nagasaki bombings. This was done under the motto "There are memories that could not be hidden away. There are memories that must not be hidden away". War experiences have to be narrated and then recorded. In 2002, the National Peace Memorial Halls for the Atomic Bomb Victims in Hiroshima and Nagasaki was founded at the Hiroshima Peace Memorial Park near the Hiroshima Peace Memorial Museum. The foundation of one more peace museum signifies the relationship between pacifism and the politics of memory. The accumulation of memories considered as "negative" should, according to this view, contribute to the growth of pacificism. The inscription of the Genbaku Dome clearly symbolizes this connection.

After the inscription of the dome, other monuments were rediscovered and recognized. The Hukuromachi Elementary School is one of them. This school is located 460 m from the hypocenter (Figure 2.8). After the bombing, the school was used as a refuge for survivors. Pupils, teachers, and local residents left their message to their families and neighbors on the burnt wall of the school with chalk. When the school was restored, a new blackboard was put on the wall, hiding the messages. In 1999, because of the deterioration of the school buildings, Hiroshima city decided to build a new structure, and during construction the messages were discovered behind the blackboard. Hiroshima city decided to preserve the messages. Specialists then deciphered the faded messages. The Hukuromachi Elementary School Peace Museum opened in 2002, and many visitors have visited since. Another example is the former Bank of Japan, Hiroshima Branch, which is also preserved and now used as an art gallery. From 2002 to 2014, it preserved the 1,000 origami cranes weighing over 10 tons and displayed part of

Figure 2.8 High-school students and volunteer guide at the hypocenter
Source: Photo by the author.

those cranes in the building. Hiroshima Army Clothing Depot is still a subject of discussion. Hiroshima prefecture decided to demolish it in 2019 for security reasons; the building could not withstand the great earthquake. But when the prefecture surveyed the citizens of Hiroshima, more than 60% of respondents were against it. The Hiroshima prefecture therefore did not take the final decision.

The case of the Genbaku Dome clearly shows a new type of cultural enhancement in Japan. The Genbaku Dome does not constitute an artistic value. In addition, it brings back bad memories. But it is considered a symbol of peace. And this evaluation is done through the social movement. It is the political, in a broad sense, that prevails over the classical conception of culture. In doing so, the concept of culture is transformed. There is no longer a clear distinction between the cultural and the political. Of course, a cultural act cannot completely exclude the political dimension. There are artists who express their political opinion in their act of cultural production. But, in the field of cultural heritage protection at least, heritage assessment should be purely cultural and academic. Culture here again means setting aside politics. But, the hidden political dimension of the cultural heritage protection until then is exposed openly with the inscription of Genbaku Dome on the World Heritage List. Cultural heritage can now be an object that reflects an idea, a philosophy, or even a political opinion if it is widely accepted in society, in this case, the idea of pacifism.

The difficulty of representation

The preservation of an object or of a building means the constitution of collective memories, but in the case of the bombing experiences, can the conservation of a building really constitute a symbol by which the survivors can remember the event? We can even ask the question if they really have to remember the experience of the bombing. On the contrary, do they not want to forget it forever? In the case of Hiroshima, which was one of the military centers in Japan, the city had been exposed to the risk of possible attack. The radioterrorism by the A-bomb targeted the entire city. It no longer distinguished between the military zone and the civilian zone. Enola Gay was indifferent to how the citizens led their daily lives even in times of war. The bombing completely changed the daily landscape. There is therefore the decisive fissure between the before and after the atomic bombardment. Under these conditions, don't the A-bomb survivors first want to rebuild their city and restore this state to the one before?

Minoru Omuta, a journalist who served as Director of the Hiroshima Peace Culture Foundation, pointed out "there are some survivors of the atomic bombing that will not talk about their experiences, no matter how often they are asked" (Ogino, 1998:202). Omuta's words recall the beginning of Claude Lanzmann's film, *Shoah*, a nine-and-a-half hour film about the Holocaust. The film begins at Chelmno in Poland, where the ruins of the concentration camp have been destroyed. Simon Srebnik, who went back to visit Chelmno, stands in a place where the ruins had been destroyed and says, "You cannot imagine the things that happened here. It's impossible. It cannot be understood by anyone. When I think about it now, even

I cannot understand it" (Takahashi 1995:152). Lanzmman evokes the impossibility of representing the event through the testimony of Srevenik.

The same type of questioning on the representation of the event can be applied to the atomic bombing. In fact, for a long time after the bombing, many A-bomb survivors were unable to talk about their experiences. There are mainly two reasons why A-bomb survivors don't want to talk about their experiences. The first reason is external and sociological. Society saw them only through the lens of their negative experience as survivors. During our investigation in 1998, around the Hiroshima area, we met a woman who quietly admitted that "while I know it is discriminatory, I would feel some uncertainty if my child told me they wanted to marry someone descended from an A-bomb survivor".

The second reason is linked to the psychology of the survivors. In order to understand it, Akiko Naono analyzes *genbaku no e* (Figure 2.9), pictures on A-bomb experiences and their authors. A TV channel, NHK, collected 2,225 pictures between 1974 and 1975 and in 2002. The Hiroshima Memorial museum keeps them and exhibits some of these pictures. Naono sees in these pictures the memory of the experience that the A-bomb survivors can never erase (Naono, 2015). She presents, for example, the testimony of Fumie Ishikawa. She drew a

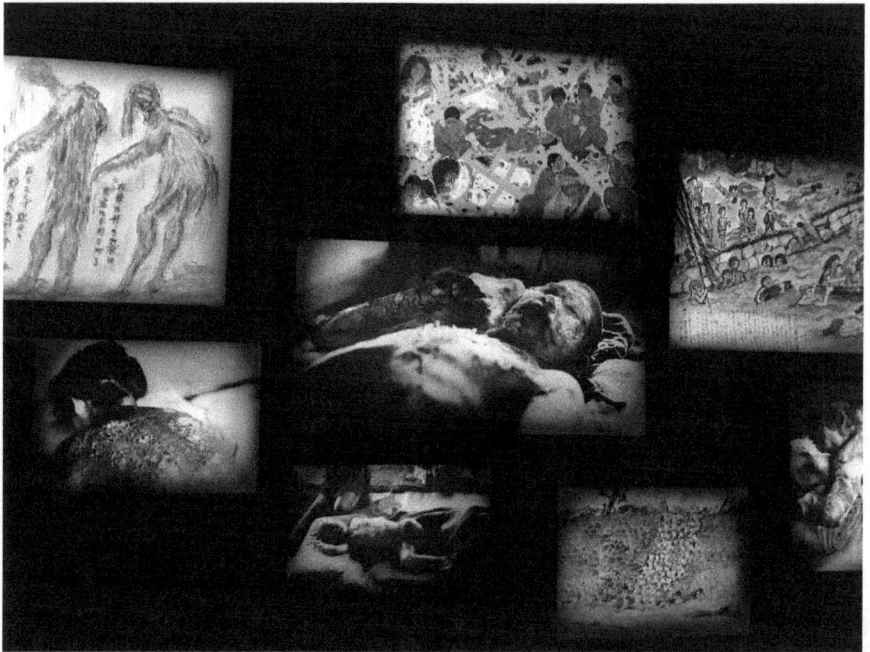

Figure 2.9 Drawings and photographs exhibited at Hiroshima Peace Memorial Museum. The drawings by Eichi Ueda and Yoshio Takahara. The photographs by Masami Onuka and Japanese Army Ship Command are part of the collection of Shogo Nagaoka

Source: Photo by the author.

scene that she saw after the bombing. She was looking for her brother at school. She saw victims lying on top of each other. It really was hell. She was dreaming of this scene and she couldn't forget it. During the interview in 2002, she kept saying, "I want to forget it". Nevertheless, Fumie Ishikawa drew and told Naono about her experiences. At the Hiroshima Memorial Museum, visitors can watch videos that collect testimonials of A-bomb survivors. It means that a number of survivors talk about their tragic experiences.

A-bomb survivors can be divided into those who remain silent and those who try to attest to their experiences. Even in an individual, there is an oscillation between the urge to tell everything and the urge to keep everything secret. This oscillation stems from the ambivalence of the world experienced by A-bomb survivors. This ambivalence comes from the split between everyday life and the memory of the world that has become hell. This nightmarish memory still lingers somewhere in their mind. If the negative memory still remains in the present almost in its original state, this situation causes precisely the trauma. There is another reason why survivors cannot easily share their experiences. It is about the difficulty of the representation. The atomic bombardment is sudden, and the victims cannot understand what happened as they cannot position themselves in the position of the observer. It is therefore extremely difficult to describe the event by remembering it. It follows that a gap always appears between what has been experienced and what is represented. Even if they want to talk about their experiences, they encounter this gap, and they hesitate to do so in the face of the impossibility they feel. The difficulty of understanding and representing experiences causes the traumatic situation (Caruth, 1995:16).

What then to do with the nightmarish past that remains in the present? Normally, time is passing, and the experiences are successively part of the past. But A-bomb experiences are still alive, and they cannot become past events. In other words, what should be in the past remains in the present. A-bomb survivors cannot escape the anguish created by their ambivalent life, and they should live in the traumatic situation. The only way to overcome this situation would be to arrange this event in the temporal order. The event must occupy a legitimate place in the regular order of time. The first solution is to pull down buildings which remind people of the nightmarish past. In this option, the disappearance of the buildings is considered as allowing the alleviation of the anxiety. Hiroshima city will then be completely innovated, and it will be like other cities in Japan. The second solution is to give them a new meaning. Hiroko Horikawa's idea goes in this direction. The Genbaku Dome will embody the experiences of the A-bomb and constitute the collective memory. The geographer Danielle Drozdzewski insists on the close relationship between place and memory. The place is not a "material backdrop", but "integral parts of the story of remembering and commemoration" (Drozdzewski et al., 2016:21). To keep the relationship between place and memory, monuments must remain in place. If the Genbaku Dome is moved and reconstructed, it will lose its authentic value. In any case, demolishing or keeping the Genbaku Dome is a question to which no one can give the right answer. All that can be said is that the citizens of Hiroshima chose the second solution, the conservation of the Genbaku Dome.

Strategy of representation

In the case of world heritage site concerning war, criterion for selection VI is important. Therefore, the representation of the events in question becomes important. The role of the museum is then increased. It gives the explanation and representation of the World Heritage site. Hiroshima Peace Memorial Museum is the pioneer of this genre. Since the 1980s, there has been a vogue for founding peace museums in Japan. The peace museum is not the army museum. It is based on the Non-Nuclear Peace declaration. A lot of cities in Japan have adopted this declaration and founded the peace museum.

But, peace museums and monuments inevitably encounter a fundamental contradiction between their tendency to generalize experiences by simplification and the will to commemorate diverse, individual experiences of war. Insofar as various experiences are increasingly generalized in the peace museum – for example by means of dioramas – this museological generalization inevitably expunges the individuality of dead soldiers for the sake of a general historical explanation and interpretation of war. The names of the fallen are erased. Their individuality is not recognized. The diversity of people's experiences is therefore lost. The Hiroshima Peace Memorial Museum, in order to resolve the contradiction, displays personal effects of the deceased, along with simple details such as the person's name and what they were doing when the atomic bomb was dropped. For example, a display of a charred tricycle includes the comment,

> Shinichi Tetsutani (3 years and 11 months) was riding his tricycle in front of his home in Higashihakushima-cho when the bomb was dropped. He suffered burns and died that evening. His father Nobuo was unable to cremate his remains, so he buried his son together with his tricycle.

Instead of simply displaying the burnt tricycle, the name of its owner was displayed. By giving this explanation, the tricycle transitioned from being an ownerless "article from the atomic bombing" to having a symbolic meaning alluding to an extraordinary event that an individual suddenly experienced. This display illustrates that the war in the twentieth century killed vast numbers of people in an instant and that nuclear weapons were the ultimate instruments for carrying out this large-scale slaughter.

The Hiroshima Peace Memorial Museum extensively renovated the central building exhibit in 2019. It pushed the pre-renovation trend, that is, respect for the individuality of the victims. It was about taking the point of view of the victims, survivors, and their families and no longer from the objectifying point of view. This is how the museum puts preserved drawings with the names of those who drew it, even their experiences. The concept of the new exhibition joins the principle in which the drawings allow to express the internal reality of the survivors. Another feature of the renovation is the exhibition of portraits of the victims. The museum exhibits objects left by victims with a short description of their life trajectory until the day of the bombardment. On the other hand, the museum has removed devices that could unnecessarily arouse fear in visitors. It is necessary to transmit the evil caused by the atomic bombardment. But the exhibition must appeal to the faculty of comprehension. Visitors are called upon to understand what really happened and place themselves as close as possible to the position of the survivors. For example, the exhibition on Sadako deals with not only Sadako but also the suffering of her parents. In a panel, it is written that financial hardships prevented from paying for painkillers and other available treatments and that their inability to do more tormented them. The exhibition, called the "N Family Breakdown", shows how a family gradually broke down. It is about the family of a fisherman N, which long suffered from the after-effects of the atomic bombardment. The family was in dire straits. N was hospitalized in 1960, 15 years after the bombardment, but was forced to leave the hospital a month later because there was nothing to do with his illness. From then on, he was completely desperate. In the last panel, it is written

> N died on January 1, 1967. Well over a hundred scars were found on his thighs. He had repeatedly cut himself with a razor during his 22-year ordeal. When he could no longer stand the agony, he distracted himself. N's death brought an end to the family breakdown.

By way of comparison, we will take the example of the Caen Memorial. The Caen Memorial shows a quite different kind of exhibition using images that illustrate the death of soldiers without recognizing their individuality. Caen is the French city largely destroyed during the Battle of Normandy and is situated near the D-Day Landing Beaches, the site of fierce fighting between the Allies and the Germans. The Caen Memorial shows, on a multi-screen display, images of Allied troops just before they landed on the D-Day beaches and of German troops preparing for the coming onslaught, both sides showing anxiety on their faces.

At first, the screen is divided into two, but the screens merge into one just at the point the Allied troops land. The Allied troops are displayed landing on the beach from a commanding aerial view, and we can see them being hit and falling one by one, as small as grains of rice. They look just like toy soldiers being pulled over by invisible pieces of thread, and in no way resemble individuals dying. This is an instance when large numbers of people are shown losing their lives instantly in a matter-of-fact fashion. In a sense, these images were taken with the supposition that they were going to die, with the objective of recording their deaths.

The D-Day Landing Beaches are part of the French World Heritage tentative list from 2014. To justify Outstanding Universal Value, it was noted that the beaches became a place of reunion for veterans, a place of transmission of the values of peace and reconciliation, compost of the Franco-German reconciliation, and they gave a lot of positive consequences for the political evolution of Europe. In 2018, the French Government submitted the file for the registration of the D-Day Landing Beaches at UNESCO.

The two peace memorials in Hiroshima and in Caen adopt a different type of display. The Hiroshima Peace Memorial Museum displays the tricycle with a caption showing the name of the owner and his destiny. On the other hand, the Caen Memorial uses footage of battles in which the individual is completely nullified. However, more than this difference, both displays give a snapshot of a situation in which individuals are crushed by an overwhelming force. These two museums provide, in this sense, two different ways to resolve the contradiction between generalization and the respect for the individuality.

Conflict of memories

It is undeniable that the Hiroshima Peace Memorial Museum and the Caen Memorial are based on the value of peace. But, as pacifism spreads into the realm of portrayal of war, it arouses feelings of resistance to this tendency. The bereaved families hope that the dead soldiers are respected and want to show that their deaths were useful for the future of society. They cannot easily accept the desire to preserve the negative heritages that pacifism promotes and hope to give a transcendental value to the death of a loved one.

Just as the Japanese Army during the Pacific War is portrayed negatively, the actions of the soldiers during the war are denied. If the actions of the Japanese military come to be considered as nothing more than atrocities committed against other Asian countries, then the sacrifice of soldiers would be forgotten forever. Veterans and bereaved families thus cannot openly commemorate their close relations or friends killed as soldiers. They feel that their desire for commemoration is suppressed because of the rise of pacifism and its dominance in peace museums.

In August 1998, Asahi Shimbun published a special issue entitled "Tormented Memories", which dealt with former soldiers' memories of war. In response to the issue, a letter to the editor voiced the opinion that "my feeling uncomfortable with this serialization is because it consciously uses interviews with people who have been through the emotional wringer of being on one hand pulled into an anything-goes war, and who have then suffered the emotional turmoil of returning to the ordinary daily life, and then ties this to pacifism" (Dated August 25, 1998). It is difficult to reconstruct the memories of such extreme battlefield circumstances at a later point in time. Even more so, it is very easy and simplistic in our society to interpret a wide range of experiences of war in support of the rising banner of "pacifism".

In fact, not everyone involved in the war has always joined peace movements. There are some people who feel that they are unable to readily align themselves with the abstract concept of peace. Some veterans and bereaved families not only have a sense of discomfort with regards to the preservation of the war heritage and the recording of testimonials to promote pacifism, but also are actually against doing so. Yasukuni Shrine[3] has, in a certain sense, become a focal point for such people.

In the Yushukan, the War Memorial Museum that manages the Yasukuni Shrine, both a *Kaiten* manned torpedo, and an *Ohka*, manned flying bomb are displayed. The US Army has granted the "courtesy" of giving the Yasukuni Shrine the Kaiten as a permanent loan, and it appears there is a special relationship between the shrine and the US military. In fact, American institutions did not adopt the absolute pacifism common to most Japanese peace museums. For example, the National Air and Space Museum in Washington, which organized the controversial exhibit of the Enola Gay that dropped the atomic bomb on Hiroshima, extols the glory of American aerospace technology and military technology. Given this, the war is, in some ways, taken as a type of game or as a sport.

Whereas the Yushukan highlights the spirit of the military, American institutions relating to war focus on the superiority of technology and skills. However, beyond this difference, the Japan War-Bereaved Association and the American Legion have many ideas in common, including that the peace we enjoy today is precisely because of the people who fought for their countries in wars. On the semicentennial of the end war, the President of American Legion sent a letter to the President of United States, expressing his opposition to the Enola Gay exhibit planned for 1995, which was to be held at the National Air and Space Museum:

> The hundreds of thousands of American boys whose lives were thus spared and who lived to celebrate the 50th anniversary of their historic achievement are, by this exhibit, now to be told their lives were purchased at the price of treachery and revenge. This is an affront to all Americans.
>
> (Nobile, 1995:xlii)

While the details of the planned exhibition did not contradict the usual American version that "the atomic bomb brought forward the end of the war, saving many lives", other viewpoints were also put forward. This meant that it was handled as a "historical debate (something regarding which there is room for discussion)" about the legitimacy of the atomic bombing. This point and the exhibit's argument that America's entry into the war spurred a feeling of revenge against Japan aggravated the American Legion. Finally, by accepting the Air Force's request, the museum abandoned the planned project and just exhibited Enola Gay without any explanation on the issue and display of the items found after the bombing. This attitude of the National Air and Space Museum ultimately influenced by the veterans opposes the Outstanding Universal Value given during the inscription on the World Heritage List: the atomic bomb is "a stark and powerful symbol of the most destructive force ever created by humankind". If so, the recourse to the atomic bomb means the most inhumane act.

At around the same time, the Yushukan held a special exhibit, "Spirits of Fallen Flowers, and a Requiem for the Truth" to commemorate the semicentennial of the war's end. This exhibit featured farewell notes from 57 soldiers killed in action, the situation of their deaths, and touched upon what sort of people they were. It is described in the catalogue of the exhibit as follows:

> Impressions from the youth who wrote down their feelings of thankfulness, realizing that their present peaceful life is thanks to the fallen who gave up their lives for their country; junior high and high school students who, in front of portraits of the fallen youth, are made aware of the preciousness of

the time that they have been given now, and who have thought of their life more seriously.

(Yasukuni jinja, 1996:211)

For a certain number of bereaved families, the Yasukuni Shrine and other shrines around Japan that venerate the war dead as "sprits of war" offer far more comfort than absolute pacifism. Let us add here that there is no mention of the atomic bombings of Hiroshima and Nagasaki in Yushukan.

Loss of negative memories

Another problem raised is that of antisepsis when monuments and objects with negative connotations are transformed into cultural heritage. The process of preservation may elide their negative aspects by the effect of antisepsis that the logic of preservation contains. Once it was inscribed on the World Heritage List, the Genbaku Dome was, to some extent, immutable. This half-destroyed ruin should be preserved in this condition to maintain its integrity. The dome is fenced, and plants are planted around it. It is harmonized perfectly with the surrounding landscape. A large number of tourists visit and take pictures of the dome, which has become an important tourist attraction.

Pacifism is another factor in antisepsis. If the Genbaku Dome is the symbol of the achievement of world peace, peace has been already realized. Hiroshima city is indeed peaceful nowadays. The dome became only a symbol of peace (Hamada, 2014:29). Complex experiences of the A-bomb tend to be reduced to pacifist ideology. At the same time, actual negative aspects are ignored. The renovation of the display at the east building of the Hiroshima Peace Memorial Museum in 2017 tends toward the same direction. The exhibition of the east building is mainly devoted to the Dangers of Nuclear Weapons, especially the movement for the abolition of nuclear weapons. But the effects of nuclear contamination have remained; *hibakusha* and their families are morally and physically suffering till today. The black rain is the radioactive fallout in the form of rain falling after the bombardment. The Japanese Government had fixed the area in which the black rain had occurred. But in fact, there were victims outside the fixed area, and they went to court. They asked to recognize that they are *hibakusha*. They won the first trial, but the government appealed a judgment. Finally, in July 2021, the victims of the black rain definitely won the case.

Nuclear colonialism at Bikini Atoll

Before concluding this chapter, we mention another World Heritage site concerning nuclear weapons. This is the case of Bikini Atoll. In 1946, the US Government decided to build a nuclear test site at Bikini Atoll in the Marshall Islands and advised residents to evacuate. Immediately after the inhabitants relocated, the US military began a nuclear test. The Bikinians first moved to Rongerik Atoll, but the life was too difficult. Then, they moved to Kiri Island from 1948. Kiri

Island was without lagoon. So, they could not continue their traditional life. Since the Bikini Atoll is no longer habitable, the city hall of Bikini Atoll is in Majuro (Figure 2.10), the capital of the Marshall Islands (Takamine, 2015).

Under these circumstances, the Republic of Marshall Islands submitted an application for World Heritage registration of Bikini Atoll in 2009. The application gives three elements of Outstanding Universal Value. First, Bikini Atoll is the testimony of "the start of the Cold War and the era of nuclear colonialism". The application proposes the concept of nuclear colonialism. Nuclear-weapon states evacuate residents from the places where they live and use them as nuclear test sites, and as a result, they have to abandon their traditional culture. Bikinians had to abandon their former canoe-based fishing activities in the lagoon and culinary tradition centered on coconut palms and breadfruits. Second, Bikini Atoll provides elements of symbolism constitutive of global culture like the mushroom cloud, the swimsuit named bikini, and Godzilla. Third, after the irradiation of The Lucky Dragon, it gives rise to the creation of anti-nuclear social movements like the establishment of *Gensuikyo*, Japan Council Against Atomic and Hydrogen Bombs (Alele Museum, 2009:52–54). The application was approved, and in 2010, Bikini Atoll became a World Heritage site.

Bikini Atoll had almost no residents except the technicians who measured the radiation dose. The nuclear test facility was abandoned. It was a place where time was frozen. Like the Genbaku Dome, it represented uncertainty. However, the registration on the World Heritage List thawed the frozen time of Bikini. Indeed,

Figure 2.10 The city hall of Bikini Atoll removed to Majuro Atoll
Source: Photo by the author.

in the application, the project to make Bikini a tourist spot for divers was mentioned. Unfortunately, it did not become a tourist spot mainly because of the lack of means of transport. It goes without saying that the World Heritage registration did not eliminate the enormous damage that the residents were suffering.

Compared to the examples treated in this chapter such as Warsaw and Hiroshima, Bikini Atoll has a peculiarity: the inhabitants could never return to their homeland. This is the fundamental tragedy of nuclear colonialism. Nevertheless, at least, the Bikinis were able to express their suffering due to nuclear colonialism and the sadness of still being nuclear nomads who are robbed of the traditional form of livelihood and forced to eat American-style canned food and drink Coca-Cola.

World Heritage and reformulation of the past

According to Barbara Adam, "the past is continuously recreated and reformulated into a different past from the standpoint of the emergent present". This possibility of interpretation presupposes a selection of events that are considered important. As a result, events regarded as unimportant become significant when a new interpretation appears. The past can be brought back and is "as hypothetical as the future (Adam, 1990:39)". Thus, there is no single, objective version of the past that history can capture. Whatever standpoint we may take, we reformulate the past based upon our present viewpoint.

The reformulation of the past does not occur without obstacles, especially with regards to controversial topics. In fact, it is difficult to give a definitive meaning to an event because it can be interpreted differently. These interpretations often enter into conflict, for they are based on different memories of the event. For example, survivors of the A-bomb and others don't recall the event similarly. Even among the survivors, there are different memories according to the circumstance in which they experienced bombardment.

The production of space is possible once conflict situations are attenuated. It goes with the regulation of different memories in conflict, and the World Heritage institutions play a role in the process of regulation and reconciliation, because the Genbaku Dome as a World Heritage site became a sacred center of Hiroshima city. Or, better yet, World Heritage institutions are implicitly mobilized for the definite reconstruction of the city.

In conclusion of this chapter, we can propose three postulates about the relationship between war and the production of space.

1   War causes not only destruction but also gives rise to the production of space.
2   To reconstruct the ruin caused by bombardment, conflict situation due to the contradiction of different memories has to be attenuated.
3   Cultural factors, including cultural heritage policy, can play a role in the process of spatial reconstruction.

Notes

1 It is noted in the leaflet of *Banteay Srey Community Tourism*.
2 The American Air Force intensively bombed Cambodian territories in 1973 to eradicate the communist party base. But these air raids produced a lot of victims, provoking hatred against the United States among Cambodian people and facilitating the rise of the Khmer Rouge.
3 Yasukuni Shrine was founded in 1869 for those who died during the internal war, and since then those who died during the war until World War II are commemorated.

3   Semantics of inscription for the World Heritage List

In the 1980s, the definition of authenticity began to be questioned. Indeed, since the inscription of Warsaw city on the World Heritage List, the conflict between the partisans for authenticity and the partisans for the revision of the classical conception of authenticity becomes explicit. Partisans for the classical concept of authenticity, including ICOMOS, are the majority. For example, one of the partisans for the revision of classical authenticity, Sophia Labadi, states that understandings of ICOMOS about authenticity do not take into account non-European approaches (Labadi and Colin, 2010:66–81). Societies which do not share the same criteria as those of UNESCO and ICOMOS would have difficulty adapting to cultural values coming from outside.

Nara Document and trends in the criteria for the selection of world heritage sites

This situation changed when the Japanese Parliament ratified the convention in 1992. Horyu-ji Temple, the oldest wooden construction, and Himeji Castle were chosen as the first candidates for World Heritage site status. But Japanese specialists had not been sure that the castle and the temple would pass the conditions of authenticity required by the Guidelines. The wooden structures like the castle and the temple are damaged faster than the stone edifice. It is thus necessary to replace the damaged parts more frequently, making it difficult to use the same material. Finally, the castle and the temple were inscribed on the World Heritage List in 1993 without opposition. Nevertheless, they felt the need to show that there was another way of heritage conservation than that practiced in Europe. Their interests joined those of European experts who intended to modify the classical principle of authenticity.

Thus, in 1994, the Nara Conference on Authenticity in Relation to the World Heritage Convention was held. Partisans for the revision of the principle of authenticity and Japanese experts took the initiative of the conference, and the Nara document was adopted. It focuses on recognizing the "diversity of world cultures and its heritages". It allows many "cultural traditions" to claim to be a World Heritage site (ICOMOS, 2020).

DOI: 10.4324/9781003014904-4

The Nara Document brings three consequences. The first consequence is the legitimization of the strategy of reproduction, which makes it possible to restore the cultural heritage in danger, and in certain cases the reproduction will substitute the original. It is thus opposed to the principle of preservation, where respect is granted to the original. The second consequence resides in the fact that the value of materiality is relativized insofar as the principle of authenticity and integrity is no longer strictly enforced. This change brings the third consequence of the Nara Conference. Regardless of whether tangible items remain, the question of whether or not to inscribe is judged on the site's cultural background. The relativization goes with the tendency of according more importance to the representation than the materiality. The discovery of cultural heritage in itself is less important than its representation and how to represent a site becomes imperative, which indicates the increased importance of the narrative in the strategy of registration.

In fact, the trends in the criteria for selection of World Heritage sites show the change in the choice of criteria. There are ten items on the list of criteria for selection of World Heritage sites: six items related to cultural heritage and four items related to natural heritage. In addition, mixed properties related to both culture and nature must satisfy definitions of both to be classified as such. Up to 2017, 867 cultural properties and mixed properties have been inscribed. Figure 3.1 shows the percentage of sites meeting which selection criteria at five-year intervals starting from 1978, when World Heritage began inscriptions. First of all, at the beginning, compared to other selection criteria, the percentage for which criterion (i) was applicable was large, but in recent years it has decreased. Selection criterion (i) is "to represent a masterpiece of human creative genius". In the initial stages, well-known places such as archaeological sites and historic buildings appreciated for their excellent architectural style were inscribed. They have always been considered world-renowned for their cultural value, even before the adoption of the World Heritage Convention. There was a common understanding of what can be considered a masterpiece.

Since diversity is valued in the Nara Document, it becomes easier to make a narrative about such a site. Selection criterion (iii) for the inscription for the World Heritage List becomes, then, more important. According to selection criterion (iii), a world heritage site is "to bear a unique or at least exceptional testimony to a cultural tradition or to a civilization which is living or which has disappeared".[1] Therefore, it is enough to demonstrate that such a site is "a unique testimony to a cultural tradition". In fact, the selection frequency of criterion (iii), high at the beginning but then dropped, has risen again since 1998. There are two types of site where criterion (iii) is selected. In the first period, archaeological sites are selected following selection criterion (iii). But, since 1998, the subject of registration is spreading to sites related to the lives of local people, not only famous ones like archaeological sites.

The increase in the number of sites registered under criterion (v) is influenced by the appearance of the cultural landscape concept proposed at the World Heritage Conference in 1992. Moreover, the global strategy adopted in 1994 pointed

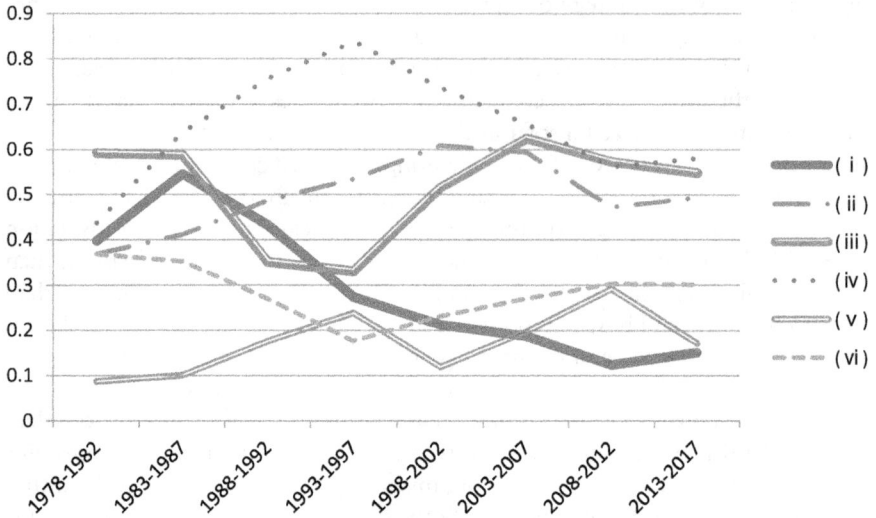

Figure 3.1 Trends in the criteria for selection of World Heritage sites. The statistics and its graphic design was done by Mayumi Yukimura

out the necessity to rectify the registration bias (UNSECO 2020j). The registration of cultural landscapes, industrial heritage, and the architectural structures of the twentieth century was newly recommended. Selection criterion (v) is to be an outstanding example of a traditional human settlement, land-use, or sea-use which is representative of a culture (or cultures), or human interaction with the environment especially when it has become vulnerable under the impact of irreversible change".[2] Regarding selection criterion (vi), according to the report of the International World Heritage Expert Meeting on criterion (vi) and associative values (Warsaw, Poland, March 28–30, 2012), "the meeting noted that in a few cases criterion (vi) may have not been recorded correctly". Furthermore, "the Committee considered that this criterion should justify inclusion in the List only in exceptional circumstances or in conjunction with other criteria". Among 13 sites inscribed on the World Heritage List only based on criteria (vi), there are certain sites related to the negative heritage. The Island of Gorée in Senegal and Forts and Castles, Volta, Geater Accra, Central and Western Regions in Ghana are directly related to the slave trade. And three sites related to the war, Auschwitz-Birkenau, the Genbaku Dome, and Old Bridge Area of the Old City of Mostal in Bosnia and Herzegovina, are registered only based on selection criterion (vi).

Semantics of inscription

The Nara document focuses on recognizing the "diversity of world cultures and its heritages". It allows many "cultural traditions" to claim to be a World Heritage site. The notion of diversity modifies thus considerably the strategy of registration, and selection criterion (iii) becomes more important. In order to show the uniqueness of a cultural heritage site, the narrative plays an important role in the strategy of registration. Therefore, local and national governments try to constitute one convenient rule for the inscription. Gradually, certain semantics of the inscription are being developed.

In Japan, two elements conduct the semantics of the inscription. First, the narrative for the inscription tends to evoke the notion of modernization and globalization. Among World Heritage sites that are registered relatively recently, at least four sites build upon this narrative based on the exchange with European civilization, which brought Christianity and industry-related ideas to science and technology. They are Iwami Ginzan Silver Mine and its Cultural Landscape, Tomioka Silk Mill and Related Sites, Sites of Japan's Meiji Industrial Revolution: Iron and Steel, Shipbuilding and Coal Mining, and Hidden Christian Sites in the Nagasaki Region. They were respectively inscribed on the World Heritage List in 2007, 2014, 2015 and 2018. Another story refers to the origin of Japanese society. The two sites represented by this story are inscribed on the World Heritage List, Ancient Tumulus Clusters in 2019 and Jomon Archeological Sites in 2021.

Especially, the Tomioka Silk Mill and the sites of Japan's Meiji Industrial Revolution are directly related to the industrial revolution and modernization.

> Tomioka Silk Mill and its related sites became the centre of innovation for the production of raw silk and marked Japan's entry into the modern, industrialized era, making it the world's leading exporter of raw silk, notably to Europe and the United States.
>
> (UNESCO, 2020f)

In this presentation of the site, two adjectives, "modern" and "industrialized", are key words. They show the site is valuable because it symbolizes modernization and industrialization in Japan. Another important expression is "world's leading exporter". It shows how Japan became a major player in the world economy. In particular, it emphasizes Japan's relationship with Europe and the United States. In the case of Tomioka Silk Mill, the exchange between France and Japan is particularly important. "The main buildings of Tomioka Silk Mill are those from the time of establishment, constructed between 1872 to 1875 that depict the technological exchange from France and Japan" (UNESCO, 2020g:129). We can often find the words "international exchange", for example, "international exchange and technological innovation in sericulture and silk-reeling". Not a simple importation of foreign technology, but exchange with European countries or reciprocity is highlighted.

A characteristic of Tomioka Silk Mill and its related sites resides in "its related sites". There are three related sites, successively, an experimental farm for production of cocoons, a school for the dissemination of sericulture knowledge and a cold-storage facility for silkworm eggs. In a way, these sites became components of World Heritage by chance.

The first site is an experimental farm for the production of cocoons. This is the residence of Yahei Tajima, who perfected a method of cocoon production. The house is not so attractive but is equipped with a thermal-powered system for temperature and humidity control named *Yagura*. When we visited the site in 2014, there were a few tourists despite it being a weekend. In front of Yahei's house (Figure 3.2), residents of the neighborhood had gathered and were talking while eating brined vegetables. The inscription on the World Heritage List made Yahei's house a neighborhood center. This is a consequence of the unexpected inscription. The inhabitants had ignored the value of the house. Once it was inscribed on the World Heritage List, the house became a symbol of the region.

The situation is almost the same for a school for the dissemination of sericulture knowledge founded by Chôgorô Takayama. Chôgorô transformed his residence into a silkworm rearing room and imparted sericulture training to young apprentices. Like Yahei's house, a few tourists visited the site. Some volunteers, all retired, were standing at the entrance. One of them told us about his life. He was born in the neighborhood but lived in Tokyo for his work. After retirement, he came back to his hometown. After the inscription on the World Heritage List, he recognized the value of the site and decided to remain here as a volunteer guide.

The same characteristics apply to the cold-storage facility for silkworm eggs constructed at the Arahune wind hole. But another characteristic is noteworthy.

Figure 3.2 Tajima Yahei's House
Source: Photo by the author.

It is the tendency to read meaning into nothingness or almost nothingness. Experts, local officials, and activists for the promotion of cultural heritage try to find meaning even in the smallest sample of old ruins where no such heritage, or very minimal components of heritage, exist. Precisely, buildings of the cold-storage facility no longer exist. It is just ruins. If we had not known the role it played in sericulture, we would not have recognized its worth. This example clearly states that the importance of inscription is accorded more to the representation of cultural heritage than to its materiality itself. In other words, the narrative about the monuments is more important than the monuments themselves.

The narrative about cultural heritage is more important than its materiality, and the narrative reads meaning into almost nothingness. This trend can be seen in other World Heritage sites. The most typical and curious example is the Ebisuga-hana Shipyard (Figure 3.3), a component of the sites of Japan's Meiji Industrial Revolution: Iron and Steel, Shipbuilding and Coal Mining inscribed on the World Heritage List in 2015. According to the description of UNESCO, these industrial heritage sites represent the first successful transfer of industrialization from the West to a non-Western nation. The 23 components are in 11 sites within eight discrete areas. Six of the eight areas are in the south-west of the country, with one in the central part and one in the northern part of the central island (UNESCO, 2020h). At the beginning, sites were only limited to the south-west of the country. In the final list, two components located in the northern and central parts were added. Therefore, the components of the sites have expanded throughout the archipelago.

At the Ebisugahana Shipyard, two Western-style ships were constructed in 1857 and 1860. But today, nothing remains on the surface, because the site is "underground" and so, the shipyard is invisible. It has not been preserved in an integral way. So, it seems to oppose the Guidelines according to which integrity is a prerequisite. But if the narrative is convincing enough, or better, deemed sufficient, a cultural property of nothingness can become a World Heritage site. Hagi city established a plan for substituting the invisible shipyard. The city will put planner markers indicating the location and the scale of the remains. Another option is to use a three-dimensional method.

Four other component sites in Hagi city are characterized by the primacy of narrative on materiality. Hagi Reverberatory Furnace (Figure 3.4) was constructed in 1856 by Chôshû han. Chôshû han was one of feudal domains of which Young Samurai would overthrow the shogunate and found the new government during the Meiji Restoration in 1868. The reverberatory furnace was just for experimentation. Hagi and other cities in the south-eastern area of Japan started the Modern Industry Heritage Project for World Heritage listing and selected this reverberatory furnace as a component of the sites in 2006. It was almost unknown to the public then. According to the curator of the Hagi Museum, at the time there was hardly any reliable historical research on the said reverberatory furnace; he did not even know who built the reverberatory furnace and when it was built. Its authenticity has not been verified (Dôsako, 2009:1). This shows the arbitrariness of the choice. It was as if the reverberatory furnace had been found by chance. Indeed, the furnace is located on a small hill behind

Figure 3.3 Invisible Ebisugahana Shipyard
Source: Photo by author.

a convenience store parking lot. Almost no one visited it before the registration. The situation changed after registration. The place has been well laid out. Groups of tourists arrive regularly. A volunteer guide told us that he had not known the existence of reverberatory furnace.

It is the same for Ohitayama Tatara Iron Works. They are located 23 km from Hagi city. They show a traditional Japanese method of making iron; a pair of bellows was used to feed air into the irons and charcoal in the furnace. This method of iron-making is known as *tatara*. Ohitayama Tatara Iron Works (Figure 3.5) operated three times; from 1751 to 1764, 1812 to 1822, and 1855 to 1867 (Hagi city, 2020). They supplied nails, fittings, and anchors to the Ebisugahana Shipyard. Even before the registration, no one visited the remains. After the registration, the toilet room has been built, and the remains have been fitted out.

Two other sites are, respectively, Hagi Castle Town and Shokasonjuku Academy. Hagi Castle town retains traditional houses and recalls the urban landscape of the Edo period (1600–1868). But it is not directly linked to the Meiji Industrial Revolution. Shokasonjuku Academy was a private school. According to the explanation of Hagi Town, "many talented graduates who played important roles in the industrialization of Japan studied at this academy" (Hagi city, 2020). The house is small and does not have value in itself. But the story of Shôin Yoshida, founder of this school, and talented graduates give importance to the house. One of the graduates, Takayoshi Kido, developed the Ebisugahana Shipyard, and he played a very important role in the realization of the Meiji Restoration. Shôin Yoshida only taught there for one year. He was the specialist in Confucian studies.

Figure 3.4 A group of tourists go to the Reverberatory Furnace from the parking lot

Source: Photo by the author.

Figure 3.5 A drawing shows the monument in its original state on the panel at the remnant of Ohitayama Tatara Iron Works. A certain imagination is required to "see" the site

Source: Photo by the author.

Shocked by the arrival of Perry's American expedition by the ship of war in 1853, he was prompted to have ideas about modernization in Japan. But his influences had to be limited. In any case, Shokasonjuku Academy was not the matrix of the ideas that led to the industrial revolution.

Officially, the nomination file is made up so as to demonstrate the originality of the Japanese model of modernization. But the components of the sites are chosen in a somewhat arbitrary fashion. In other words, to show the story on the "original" method of Japanese modernization, Hagi city has gathered here and there elements that could contribute to the constitution of the story. Why is it a bit arbitrary? This is because the explanation of industrialization puts too much emphasis on the political role played by a feudal clan. Indeed, the components of the Hagi's sites are not directly related to the industrial revolution proper compared to the components of the Imperial Steel Works in Fukuoka founded in 1901.

Narrative and conflict

The primacy of the narrative in World Heritage institutions gives rise to various conflicts. The first conflict aroused is due to the process of preparing the nomination file. The choice of site components is negotiated. In particular, from the moment the serial nomination[3] is suggested, local communities seek to make their "heritage" components of the sites. It is therefore necessary to constantly adjust the list of components to the driving story on the sites. During the preparation of the nomination file, the story is constituted little by little and sometimes largely modified. Therefore, depending on the story, the components of the sites change.

In 2006, an application of "The Modern Industrial Heritage Sites in Kyushu and Yamaguchi" was submitted to the Agency for Cultural Affairs. This was the first official attempt at World Heritage listing on the matter. In the first application, only the Reverberatory Furnace was selected with regard to the components related to Hagi city. Ebisugahana Shipyard was added when "The Modern Industrial Heritage Sites in Kyushu and Yamaguchi" were listed to the World Heritage Tentative List of UNESCO in January 2009; a Tentative List is an inventory of those properties which each state party intends to consider for nomination. After that, the Agency for Cultural Affairs then asked to accomplish two objectives. It was about to prove that the sites showed the model of modernization in Asia and that modernization in Japan stemmed from Japanese artisanal tradition. Taking into account these suggestions, the Advisory Committee, established under the Consortium for the Promotion of the Modern Industrial Heritage to inscription on the World Heritage,[4] drafted the third version of the list in October 2009 to which Hagi castle town and Ohitayama Tatara Iron Works were added. It is considered that the craftsmen of the Edo period and the traditional method of steelmaking helped to implant modern technology. And it was in the final version of the list in 2013 that Shokasonjuku Academy was definitely added to (Kimura, 2014:222–224). On the contrary, some monuments are removed from the list. For example, Ita mine shaft oar in Tagawa city, constructed in 1910, was eliminated from the October 2009 list, because it didn't fit well with the modernization narrative in Japan.

In the negotiation process, expert opinion is important and in principle the Culture Council that belongs to the Agency for Cultural Affairs should play a major role. But, in the case of modern heritage, the opinion of experts is not always respected. This was the case for the Genbaku Dome, in which the civil movement in Hiroshima played a major role. In this kind of heritage, the selection of components is no longer the monopoly of the Culture Council made up of experts in "traditional" heritage such as temples. The appearance of new actors in the field of cultural heritage necessarily provokes conflicts between traditional experts and these new actors.

The case of the Sites of Japan's Meiji Industrial Revolution is a good illustration of the conflict between old and new actors. In 2006, the Ministry of Economy and Industry organized a colloquium on industrial heritage and world heritage and since then has pursued the possibility of inscribing the heritage of the industrial revolution on the World Heritage List independently of the Agency for Cultural Affairs. The Ministry of Economy and Industry placed the inscription on the World Heritage List within the framework of the development of local cultural resources. Therefore, it sought to find a way to protect active industrial heritage that the Law for the Protection of Cultural Property could not cover. In doing so, the Ministry of Economy and Industry, backed by the business community, was seeking to pave the way for inscription on the World Heritage List. However, the Cultural Council and ICOMOS Japan were not really in favor of the same. There was therefore tension between the Ministry of the Economy and Industry and the Agency for Cultural Affairs.

In 2012, the Cabinet Secretariat took charge of the registration process whereas the Agency for Cultural Affairs generally does. If this is so, it is because among the component sites, there are some that are active according to the explanation of the Cabinet Secretariat. For example, the Mitsubishi Shipyard in Nagasaki (Figure 3.6) was in activity. The Agency for Cultural Affairs could not deal with them. This is how the Department of the Industrial Heritage WH inscription of the Japanese Cabinet Secretariat was created. It brought together representatives of several ministries that include the Ministry of Foreign Affairs, the Ministry of Economy and Industry, the Ministry of Land, Infrastructure, Transport and Tourism, and finally the Agency for Cultural affairs.

In January 2014, to submit the candidate to the World Heritage Commission, there was a struggle between the Meiji Industrial Revolution Sites supported by the Cabinet Secretariat and the Hidden Christian Sites in the Nagasaki Region supported by the Agency for Cultural Affairs. The Cabinet Secretary of the government of Shinzo Abe made the final decision and obviously chose the Meiji Industrial Revolution Sites proposal. Two former officials from the Ministry of Education and Science complained that the government failed to respect the Cultural Council's decision. They pointed out, in *Shūkan Asahi*, dated June 23, 2017, that from the start, Prime Minister Abe intervened in one way or another in this affair through the cabinet secretariat. In fact, during the colloquium of 2006, Abe's wife spoke as a "supporter;" Abe was then cabinet secretary. It can be noted that the prime minister comes from Yamaguchi prefecture including Hagi

Figure 3.6 A former factory building of the Mitsubishi Shipyard today used for the
museum of the shipyard, a component of the World Heritage site
Source: Photo by the author.

city. One of the former officials interviewed in the article pointed out that it was
strange to locate the beginning of industrialization in the establishment of Shokasonjuku Academy in Hagi city. That said, the government was in the hands of the
left-leaning Democratic Party between 2009 and 2012. The Liberal Democratic
Party, a conservative party to which Abe is a member, was in opposition at the
time and only in December 2012 did the Liberal Democratic Party return to
power. And Abe was again the prime minister; he was the prime minister for the
first time between October 2006 and September 2007. Therefore, at least during
the reign of the Democratic Party, Abe could not get a grip on decision-making.

In this example, there is another type of conflict. This is the conflict between
the traditional experts supported by the Agency for Cultural Affairs and the
newcomers in the domain of cultural heritage supported by the Ministry of the
Economy and Industry and the business world. It's a bit like the symbolic struggle described by Pierre Bourdieu in the field of power (Bourdieu, 1979). The
traditional experts come from the field in which cultural capital dominates while
the newcomers find themselves in the field in which economic capital prevails.
The conflict arises between the traditional experts and the newcomers who have
different narratives to promote a site.

The choice of sites for Meiji's Industrial Revolution seems to signify victory for the business camp. In fact, things are more complicated insofar as there are different types of movement for the promotion of industrial heritage. This is the case with Hashima Island (Figure 3.7), generally called Gunkanjima, which means battleship island. The Hashima coal mine is situated on this island near Nagasaki city, which endured the atomic bomb. Excursion boats regularly go to the island so that tourists can visit limited areas. Volunteers guide visitors and explain the history of the Hashima coal mine. Their narrative also tends to emphasize the glory of the mine. The narrative evokes only positive Outstanding Universal Value by arousing a feeling of nostalgia toward the time of glory.

In this case, the movement begins with the idea of the ancient inhabitants of the island. They created the Association to Make Gunkanjima a World Heritage site. The founders of the association come from the families of miners. The president of the association, Dôtoku Sakamoto, got the idea to make this island a World Heritage site when he organized a reunion with his former classmates from Hashima High School on the island itself in 1999. This was the emergence in him of museological desire. At this point, he felt compelled to retain his familiar place where he spent his childhood as it was forbidden to land on the island after the closure of the Hashima coal mine in 1974. The approach is very different from that of Hagi city. It is not the municipality or politicians, but the voluntary association of former inhabitants that has started to promote Hashima island as a heritage site (Sakamoto and Goto, 2005). Sakamoto wrote in his book released in 2005 that Gunkanjima as a World Heritage site was a gigantic dream. When he told his dream precisely, many said that it was impossible to achieve it. The movement that was created had, at the beginning, nothing to do with the

Figure 3.7 Hashima Island
Source: Photo by the author.

municipalities which showed no interest in the project. In particular, the former inhabitants of Hashima Island were against it. But the movement joined the wider movement for the promotion of industrial heritage supported by local and central government politicians and the business community.

The movement for the inclusion of Industrial Revolution Sites on the World Heritage List is therefore heterogeneous. There is a sort of nostalgic nationalism on the part of the conservative political milieu that places too much emphasis on the role played in modernization by the young samurai of the southern region and Shôin Yoshida allegedly ideological leader of the Meiji Restoration. There is also the feeling of nostalgia for the near past by the miners and their families who have been forced to permanently leave their homes due to the closure of the mines. The Sites of the Meiji Industrial Revolution encompass all of these varied ideologies and sentiments. These nostalgic ideologies arouse les effects of antisepsis. They glorify the past when the mine was still actively producing coal, without mentioning negative aspects of the mine. The site has thus been completely sanitized physically and symbolically. Negative memories seem to have been almost definitively eradicated.

Conflict arising from the memories of colonization

One element that worried the promoters of The Sites of Meiji Industrial Revolution was the issue of forced labor to Koreans and Chinese. Thus, the nomination file considers the fact that the industrial revolution begins in Japan from the 1850s and ends before 1910. The year 1910 is the year of the annexation of Korea. After the annexation, many Koreans worked in the coal mines, especially during the wartime. If the narrative is limited to before 1910, the question of the treatment of Korean miners does not need to be raised. But, contrary to what was expected, the inclusion of mining establishments on the World Heritage List has caused disagreements with South Korea.

Just prior to the commencement of the discussion on the Meiji Industrial Revolution Sites at the World Heritage Committee, South Korea gave an adverse opinion to Japan's proposal, for there was no explanation regarding forced workers in the mines. The German delegation intervened in this quarrel and sought mutual understanding between Japan and South Korea. Thus, the Japanese delegation announced the following statement.

> Japan is prepared to take measures that allow an understanding that there were a large number of Koreans and others who were brought against their will and forced to work under harsh conditions in the 1940s at some of the sites, and that, during World War II, the Government of Japan also implemented its policy of requisition.
>
> (UNESCO, 2015:222)

The Korean delegation responded with the following statement.

> The Government of the Republic of Korea has decided to join the Committee's consensus decision on this matter, as it has full confidence in the

authority of the Committee and trusts that the Government of Japan will implement in good faith the measures it has announced before this august body today.

(UNESCO, 2015:223)

Japan therefore took responsibility for taking action on what it had promised at the World Heritage Committee meeting. So, the Japanese Government opened the Industrial Heritage Information Center in 2020. This center belongs to the Cabinet Secretariat and not to the Agency for Cultural Affairs. As soon as it is opened, South Korea criticized the information center's exhibit in which there was really no description on forced laborers. But the center ignored this criticism. The Korean Government issued the statement against the information center exhibition. According to this statement, the Japanese Government did not at all respect the agreement reached at the meeting of the World Heritage Committee. The Japanese Government responded by saying that it did not intend to change the center's exhibition. The center chose the testimony of Fumio Suzuki, born in Hashima in 1933, whose parents were Korean and his father was middle manager of the Hashima coal mine and was not a forced laborer. He remained in Hashima until 1942. He said there was no difference between the Japanese and the Koreans. But Suzuki's testimony was only in childhood memory. And since the Japanese and the Chinese and Korean workers did not live in the same apartment, most Japanese residents of the island could not really observe the living conditions of the forced laborers. In addition, the mobilization of Korean workers began from 1939 and accelerated from 1941 for lack of Japanese laborers. However, Suzuki left the island in 1942 and did not know what happened afterwards, especially after 1944, when forced labor was officially applied to Koreans.[5] On the other hand, the center interviewed only former residents of the island and not Korean ex-workers. Moreover, the information center did not consult testimonies already collected by researchers.

Thus, the conflicting situation over the Hashima coal mine between the Japanese Government and the Korean Government reappears. This conflictual situation is not only happening at the government level. Indeed, on the Japanese side, the nostalgia for life more or less embellished on the island is underlined. It dismisses its negative side and values the community relations that would have existed on the island. On the Korean side, life on the island doesn't mean a nostalgic feeling at all. On the contrary, it was a nightmare. The number of Koreans who worked in Hashima is estimated to be around 750 people (Nagasaki zainichi chosenjin no jinken o mamorukai, 2016:161). There have been testimonies of the extremely harsh life on the island, the lynching carried out in the mine tunnel, etc. Note that in this controversy, the Mitsubishi Nagasaki Shipyard, another component of the Sites of the Industrial Revolution in Nagasaki, was not put on the agenda. There were also Korean and Chinese workers in this shipyard. In addition, the shipyard was affected by the atomic bombing on August 9, 1945. The same is true for the Miyanohara pit at the Miike Coal Mine in Fukuoka, one of the biggest coal mines in Japan. Many Chinese and Korean laborers worked there and were sometimes victims of explosions in the mining gallery.[6]

The case of Hashima shows explicitly that memories of colonization provoke conflicts. Two opposite interpretations of the past appear, and they are hardly reconciled. It is partly since there are not enough written documents, which means that in order to reconstitute the historical reality, it is necessary to have recourse to the witnesses concerned, in other words, to various individual memories. They necessarily change over time, and different memories enter contradiction between them. Furthermore, the conflict may become explicitly political to the extent that governments intervene directly in the matter. The inscription of a site on the World Heritage List can therefore give rise to conflicts that the World Heritage Committee cannot always resolve.

So far, we have shown narratives about two sites that have been on the World Heritage. We can now summarize what we have observed. First, both are linked precisely to modernization in Japan. The story for the nomination file draws inspiration from this moment of great change in the nineteenth century. The idea of considering Japanese modernization as original thus becomes the source for designating monuments as cultural heritage. The narrative must be built around the modernization in Japan and its originality. Second, at first glance, these narratives seem to be "diverse". But in fact, they have a certain common grammar. Sites likely to correspond to the grammar of World Heritage listing are then sought and chosen. Third, the narrative that shows the originality of Japanese modernization implicitly attaches itself to a kind of cultural nationalism. And this soft nationalism makes use of the World Heritage institutions.

Narrative about globalization and Japan

The two sites presented earlier show in part the contact with the West and its influence of modern technology. The narrative of two other sites describes the relationship with Europe in a different way. They are Iwami Ginzan Siver Mine and its cultural landscape and Hidden Christian Sites in the Nagasaki Region. The first is still the mine, but that has little to do with modern industrialization. In the description of Outstanding Universal Value, it is presented as follows: during the Age of Discovery, in the sixteenth and early seventeenth centuries, the large production of silver by the Iwami Ginzan Silver Mine resulted in significant commercial and cultural exchanges between Japan and the trading countries of East Asia and Europe. In comparison with other mines like Osarizawa mine in the northern part of Japan, the gallery of which is open to the public and easy to access,[7] Iwami mine is open to the public only after the mine has been inscribed on the World Heritage List. Other components like the Ōmori area, mining town formed in seventeenth century, or Yunotsu (Figure 3.8), port town famous for hot springs are not so outstanding by themselves. Both retain old quarters that date back to when the mine was prosperous. But there are many other places in Japan that retain old houses and buildings. Inscription on the World Heritage List adds additional value to these neighborhoods. But, in the case of Yunotsu, the village does not seem particularly interested in advertising with World Heritage. It is already known for its spa and attracts tourists.

Figure 3.8 Yunotsu town
Source: Photo by the author.

Another site is Hidden Christian Sites in the Nagasaki Region. Christianity was brought to Japan in 1549 and spread to the region of southern Japan, particularly to Nagasaki. The Tokugawa clan shogunate, which ruled Japan from 1603, banned Christianity in 1614. Some Christians continued to be so despite the strict ban. Some of them revolted against the shogunate; the Shimabara Rebellion occurred in 1637 in Nagasaki. Christians who wanted to keep their faith fled to the islands, which are difficult to access from the outside, and continued to be Christians. The existence of these hidden Christians is described in Shusaku Endo's novel, *Silence*, from which Martin Scorsese made a film. The Meiji government secured freedom of belief to the hidden Christians in 1873 and the Christians finally openly built their churches. Initially, the Nagasaki prefecture made efforts to make them components of the site. But they are not monuments built in the time of the hidden Christians. Thus, following the suggestion of ICOMOS, the villages where the hidden Christians led their life became components of the sites. For example, in the Gorin Village on Kuga Island, the former Gorin Church (Figure 3.9) was built in 1881 after the acquisition of freedom. It is not the church but the isolated village, which has little land and is difficult to access, that is a component of the site. It recalls the time of the hiding. Nothing

Figure 3.9 Gorin village and Former Gorin Church in Kuga Island
Source: Photo by the author.

of the time of the hidden Christians remains. Only memory is vivified. The history of the hidden Christians is that of contact, of the break, and the contact again with Catholicism. The passage from the period of the rupture to the period of the contact again is important insofar as the contact again implies the freedom of expression with the acceptance of the religion until then prohibited.

Origin of Japanese society

Another narrative for registration refers to the origin of Japanese society. The first case is Mozu-Furuichi Kofun Group: Mounded Tombs of Ancient Japan, which was inscribed on the World Heritage List in 2019. The second case is Jomon Archeological Sites registered for the World Heritage List two years after.

First of all, what is the Mozu-Furuichi Kofun group? To know it, we have to go back to the nineteenth century. The intellectuals and officials of the shogunate began to identify the ancient burial mounds from 1850s. Each ancient burial mound is attributed to an emperor, *Tenno*, even though there was no evidence. So, there may be errors, and in any case the word *Tenno* did not exist in the fifth century, when the ancient burial mounds were built; the great king was called *Ôkimi*, and it is not obvious whether these *Ôkimi* were ancestors of the imperial family. This rediscovery of the ancient burial mounds in the nineteenth century is closely linked to the rise of the imperial ideology that would establish the Meiji government that had, thus, a strong interest in legitimizing the genealogy of the imperial family and used ancient burial mounds to justify the ideology of a unique genealogy.[8]

After the defeat at the Pacific War in 1945, the imperial ideology was denied, and the status of the emperor changed. The Imperial Household Agency, the former Imperial Household Ministry, which deals with the imperial family's affairs, implicitly sticks to the mythology of the past. And for this reason, the Agency refused to open ancient burial mounds to the public. On the contrary, archaeologists recognize the need to excavate ancient burial mounds to obtain knowledge about the time they were built. The agency and archaeologists therefore do not have the same opinion on ancient burial mounds. To begin with, archaeologists and the Imperial Household Agency do not have the same terminology for ancient burial mounds. Some *kofun*, which means ancient burial mounds, are called *ryobo*, especially by the Imperial Household Agency, because they are considered *ryobo* imperial mausoleums. On the contrary, archaeologists suggest the use of the term *kofun* for all mounds and to avoid the term *ryobo*. Therefore, some mounds have two different appellations. For example, the largest mound of Mozu-Furuichi Kofun Group is called *Daisen kofun* by archaeologists and historians and Nintoku Tenno *Ryo* by the Imperial Household Agency; Nintoku Tenno is the name of an emperor recorded in *Kojiki*, the oldest Japanese chronicle edited in the eighteenth century. *Ryo* has the same meaning as *ryobo*. Naofumi Kishimoto, an archeologist and an expert on the subject, advanced the argument that this mound should not be attributed to Nintoku Tenno but to another one (Kishimoto, 2017:77).

Archaeologists asked the Imperial Household Agency to conduct research inside ancient burial mounds from 1976. Since 1978, the agency allows specialists to conduct a survey on one ancient burial mound every year, but the specialists have access only to a part of the ancient burial mound on which they conduct research. This is how archaeologists started the movement for inscription on the World Heritage List. This attempt was for archaeologists a research strategy. If the inscription on the World Heritage List is achieved, it is because the Imperial Household Agency has granted a greater degree of freedom to archaeologists. That being said, the same situation continues after the inscription on the World Heritage List. The Agency for Cultural Affairs, normally in charge of the World Heritage sites, agrees that the Imperial Household Agency continues to assume the same responsibility for the ancient burial mounds after the inscription. Therefore, archaeologists fear that the imperial agency will continue to keep the same closed attitude, because the agency was not in favor of the inscription on the World Heritage List while the archaeologists pursued it. Indeed, this should allow them to freely research the ancient burial mounds.

The narrative about Jomon Archeological Sites is also related to the origin of the Japanese society, but it depicts another type of story. The representative of this story is a philosopher, Takeshi Umehara. According to Umehara, the

Sannai-Maruyama ruins prove his hypothesis that the foundations of Japanese culture lie in the Jomon culture. Umehara noted,

> The Sannai-Maruyama site shows that in the Jomon era (about 16,000 BC to 3000 BC), a very high-quality civilization with abundant food production existed. And a series of important discoveries in this site, reveal the mentality of the Japanese.
>
> (Umehara et al., 1996:18)

In fact, local residents had known the existence of the ruins, and some residents had collected earthenware pieces which had retained their integrity. But when the construction of a baseball stadium began in 1992, it was discovered that the site was much larger than had been previously believed. After two years of research, in 1994, Aomori prefecture decided to cancel the construction project to preserve the site. Since then, the research on the site continues, and the influence of Umehara's ideology seems attenuated. Nevertheless, his idea of Jomon culture as foundation of Japanese culture propelled the excavation of the ruins. Graves were discovered, and archeologists are beginning to assert that the site shows the beginning of the sedentary life and the emergence of villages. These findings could prove the existence of "the first civilization in Japan".

Political aspects of inscription

The process of inscription on the World Heritage List presupposes a lengthy negotiation. The negotiation continues even after the government has chosen a candidate for the World Heritage List. In the first phase, the nomination file is evaluated by ICOMOS dedicated to the conservation of the world's monuments and sites in accordance with the Operational Guidelines for the Implementation of the World Heritage Convention. In the second and last phase, the World Heritage Committee holds an ordinary session once a year and makes a final decision.

The uncertain situation continues throughout the negotiation and it sometimes provokes conflicts. As mentioned earlier, there was the conflict between the different ministries in the case of Sites of Japan's Meiji Industrial Revolution before the submission of the nomination file. Even after the submission, those who promote registration sometimes encounter obstacles and are involved in conflicting situations. They are therefore obliged to face it until the last moment. The first conflict is between experts and non-experts. Normally, experts possess the power of decision-making. Indeed, according to the Rule of Procedure, "States members of the Committee shall choose as their representatives persons qualified in the field of cultural or natural heritage". However, the Convention of the World Heritage Committee's 35th session designates that "the delegations are directed by diplomats who less and less frequently call upon their experts" (UNESCO, 2020k). In fact, the members of the Committee include many diplomats. It shows explicitly that decision-making depends not only on the academic point of view but also on diplomatic negotiation. This means non-experts who represent the interests of their country intervene. Experts claim to respect the value they consider universal, while diplomats try to impose national interests.

The second conflict arises among the experts insofar as not all experts share the same principle of heritage conservation. The question of materiality or authenticity still provokes controversy among experts. It stems from a fundamental difference in what is true cultural value. Although they remain indispensable conditions of cultural value in World Heritage institutions, the value of authenticity and integrity is nowadays undermined. The third conflict takes places between different national delegations. In some cases, explicit political conflicts take place. In other cases, the conflict is more nuanced by taking a purely cultural aspect. There are societies where materialistic value is dominant; materialistic value privileges precisely the material integrity of objects and monuments. There are others that have the constructivist conception of cultural heritage; a heritage is not always there as materialists want to believe, but it appears only when it is represented as such.

The case of the Fujisan site is a good illustration of all these three types of conflicts. A Japanese diplomat, Seiichi Kondô, describes it through his experience of negotiation at the World Heritage Committee held in 2013 (Kondô, 2014). The Japanese Government submitted the nomination file of Mount Fuji (Figure 3.10), Fujisan, the previous year. Mount Fuji is a dormant volcano and the highest mountain in Japan. The title of the nomination file for the World Heritage List is "Fujisan, sacred place and source of artistic inspiration". The site is composed of 25 components, including the Miho no Matsubara pine tree grove. It is located on a beach where travelers could view and appreciate Mount Fuji. This landscape attracted, in the nineteenth century, famous artists such as Hokusai Katsushika (1760–1848), a representative of ukiyo-e, Japanese prints, who drew Mount Fuji, the sea face, and the pine tree grove.

Hokusai renovated the technique of perspective introduced by the Akita school. Under the influence of Western painting brought by the Dutch, a school of Western painting developed in Akita, in the northern part of Japan. According to the interpretation of art historians (Inaga, 1983:29–45), this Akita school, by transforming the linear perspective imported from Europe, created a new conception of space in art. It is a conception of heterogeneous space; what is near coexists in a single image with what is far. The close-up is superimposed on the background without the intervention of the intermediate second shot. For example, in "Shinobazu Pond", Naotake Odano puts a potted flower and the pond in a painting like a composite photograph. The potted flower presents the proximity, the pond, the distance.

Hokusai seems to have followed the path pioneered by Akita school. But in fact, he brought about a big renovation in the way of treating the perspective. In his works, the image is broken down into three parts and no longer into two. A series of prints by "Thirty-six Views of Mount Fuji" (1823–1833) illustrates this three-part composition well. Hokusai describes 36 Mount Fuji views seen from 36 different places. Regarding Mount Fuji, Kôkan Shiba, one of the pioneers of Japanese perspective, remarked that Mount Fuji rose alone, it was better to look at from far. For example, "Hodogaya" is made up of three parts; travelers

Figure 3.10 Mount Fuji
Source: Photo by the author.

in the foreground, Mount Fuji in the background and between the two, the land in brown and the forest in green. In this composition, the subject of the print, Mount Fuji, is in the background and not in the foreground. In the case of Shinobazu Pond, the background is still blurry. But, in Hodogaya, Mount Fuji, drawn in the background, is well focused. Another original element of its composition is the presence of the intermediate part between the foreground and Mount Fuji. In the case of Hodogaya, it is the unremarkable brown-colored part that shows the distance between travelers and Mount Fuji. The same composition is observed in "Gohyakurakanji Sazaedô". Travelers looking at Mount Fuji from a temple, Mount Fuji and the land, always colored brown, are drawn.

If he found this original composition, it is because he placed himself in the position of the traveler. He himself moved to 36 different places that would allow him to have a good look at Mount Fuji, and he drew his 36 views. His works therefore reflect the gaze of the traveler or the stranger on the landscape. The landscape is something the traveler looks at and sometimes enjoys. He therefore has an objectifying gaze while peasants, fishermen, etc., do not regard their environment as the traveler-stranger. Hokusai describes this difference well. In these works, travelers enjoy the landscape whose center is Mount Fuji. On the other hand, those who work are not interested in their daily space from an aesthetic point of view. They don't have enough time to appreciate the beauty of Mount Fuji. For example, the fisherman throws his fishing net into the rough sea, but he is not looking at Mount Fuji. Thus, Hokusai expresses the emergence of mobility that the development of the market economy has involved and the resulting new conception of space. This conception constitutes the gaze which goes beyond everyday space and awaits a transcendent existence. Mount Fuji is described aesthetically as a transcendental existence. Hokusai was well aware of the expansion of the market economy. In "Mitsui's Shop in Suruga District", three workers work on the roof of Mitsui's house, and, as always, Mount Fuji is in the background. In this print, two signs of the Mitsui house, on which are brands, "Fixed prices" and "Payable in cash" are noticed. Indeed, he himself sold his works on the market of prints, which was, at the time, very popular. A new type of aesthetic gaze, independent of religious value, was born in this period.

In spite of all this history that proved the close relationship between Mount Fuji and Miho no Matsubara (Figure 3.11), ICOMOS suggested excluding Miho no Matsubara from the components of the World Heritage site, because Miho no Matsubara is not part of the Fujisan region. Faced with this decision, the Japanese delegation had two options. The first was to accept the suggestions of ICOMOS to avoid a potential source of conflict. The second option was to make the World Heritage Committee understand the value of the pine grove so that it could be included in the components. Kondô chose the second option and tried to negotiate with the members of the World Heritage Committee, some of whom were diplomats he was familiar with. In his book, he highlighted the hegemony of European countries in decision-making. This was the reason he made contacts mainly with representatives of four European countries. ICOMOS and one or two European delegations had a negative opinion to the listing of Miho no Matsubara because the relations between Miho no Matsubara and Fujisan could not be empirically verifiable. This controversy stems from the traditional conception of cultural heritage and the new one. For ICOMOS and some European delegations, the materiality of the heritage is essential. The Japanese delegation opposed this negative position by stressing that these relations are part of the collective memory of the Japanese (Kondô, 2014:26–27). They also emphasized the importance of storytelling about these relationships. Concretely, the Japanese delegation emphasized the "intangible links" between Miho no Matsubara and Fujisan. The intangible links mean here, the coordination of two distant places by an aesthetic value. This value develops from the eighteenth century and it expresses, in particular, in ukiyo-e, the new conception of space which is closely linked to social change occurred at the time.

In this context, the inscription of Miho no Matsubara on the World Heritage List was an attempt to link the two different sets of logic: that of World Heritage institutions and that of Japanese delegation. To abandon the inscription of Miho no Matsubara was to accept the traditional logic of World Heritage institutions. But, the Japanese delegation dared to bring their logic into the framework of the World Heritage institutions. The materiality of the objects of the monuments is not everything. The intangible links that connect Miho no Matsubara to Mount

Figure 3.11 Miho no Matsubara without Mount Fuji not visible on a crowdy day
Source: Photo by the author.

Fuji should be an Outstanding Universal Value. Finally, Miho no Matsubara was chosen at the committee meeting as a component of the world heritage site, contrary to the recommendations of ICOMOS.

An important difference concerning the relationship of the aesthetic gaze to the object exists between classical cultural heritage institutions and the Japanese delegation. In the case of classical cultural heritage institutions, the object is more important than the aesthetic gaze of the person who appreciates it. It is therefore from the object that the institutions of cultural heritage are constituted. On the contrary, the Japanese delegation tried to objectify the aesthetic gaze, which is itself intangible.

Illusion of diversity

The example of Fujisan shows how difficult it is to gain acceptance for a concept of cultural heritage that has not yet taken its place in the existing standards of World Heritage institutions.

The Nara Document introduced the concept of cultural diversity which is expected to open up different types and ways of preservation of cultural heritage. However, the concept of diversity cannot always marry up with culture that has no tradition of preservation of cultural heritage publicly. In other words, the problem with the concept of diversity is the contradiction between diversity and universal value. The Guidelines insist on the Outstanding Universal Value, which is transcendental for all humanity. As such, World Heritage sites should be passed on to future generations. But, in fact, what does "all humanity" mean? The concept of all humanity presupposes the existence of a transcendental value system that the international community can share as a whole. Does this value system, common to all people, really exist? Do all societies share the same value? And, in particular, does not the concept of universal value contradict the concept of diversity insofar as diversity presupposes the right to be different? What can be pointed out about this series of questions is the following. The notion of diversity has left more freedom and as a result various types of semantics are appearing for the designation of objects and monuments as cultural heritage. This sometimes implies conflicts between new semantics and the classical norms of heritage protection. This means a new phase for World Heritage institutions, which we have already called the Age of Preservation. As already mentioned, the Age of Preservation tends to negate the classical norms of cultural heritage institutions. The introduction of the notion of diversity can be seen as an attempt of adaptation to this new phase.

Notes

1 The Nara document states, in an abstract way, that all judgments about values attributed to cultural properties as well as the credibility of related information sources may differ not only from culture to culture, but also within the same culture, and that it is thus not possible to base judgments of values and authenticity within fixed criteria. The document adds that, on the contrary, the respect due to all cultures requires that heritage properties be considered and judged within the cultural contexts to which they belong. This idea of diversity regarding cultural values stated in the Nara Document was finally expressed in the revised Guidelines of 2005.
2 UNESCO monitors the regional bias of inscribed properties and trends in selection criteria in order to maintain the balance of the area and the kind of registered properties.
3 A serial nomination is any nomination which consists of two or more unconnected areas. A single World Heritage nomination may contain a series of cultural and/or natural properties in different geographical locations, provided that they are related because they belong to: (1) the same historico-cultural group; (2) the same type of property which is characteristic of the geographical zone; or (3) the same geological, geomorphological formation, the same biogeographic province, or the same ecosystem type, and provided that it is the series, and not necessarily each of its components taken individually, which is of outstanding universal value (UNESCO, 2020i).
4 Consortium was established in October 2008.
5 But, strictly speaking, the Korean mobilization, before 1944, took on the appearance of forced labor (Mizuno, 1998).
6 There are other facts that evoke negative memories. Great labor disputes occurred in 1953 and from 1959 to 1960. The last dispute divided laborers into two opposite groups. Moreover, the great mine disaster happened in 1963. It caused 458 deaths and many victims of anthracemia. After the mine's closure, many pit workers left the area of the Miike Coal Mine to seek another job.
7 About 1.7 km of the mine shafts were opened to the general public as "Mineland Osarizawa."
8 Chapter 4 explains in detail imperial ideology.

4   Intangibles and tangibles
The logic of actualization

In Chapter 1, we pointed out that when a nation-state was being formed, the order of recollection could be a serious obstacle. The order of recollection privileges the relation of the living with their respectable existence such as their ancestors and deity. In doing so, it controls the entry of strangers into the community space. The nation-state consists in promoting free trade with the opening of society to the outside, but the order of recollection resists this opening. Therefore, promoters of the nation-state have to remove this obstacle one way or another. Overall, there are two ways to do with it. The first way is to destroy it. A typical example is the French Revolution, which denied the authority of the Church. Instead of religious authority that assured the social order of the parish, the republic should create a new foundation that could govern heterogeneous communities. This is how the museological desire appears with the capitalist desire, and the institutions of cultural heritage legitimize it. Monuments that symbolize national history become the foundation of the nation as cultural heritage. World Heritage institutions have emerged from cultural heritage institutions formed in the nineteenth century in Western Europe to create a pivot of the nation-state.

Japan did not follow the same path to constitute the institutions necessary for the formation of the nation-state, because Japan chose the second way to do with the order of recollection. In fact, Japan did not adopt the same principle vis-à-vis the order of recollection as Western Europe. Instead of destroying the order of recollection, the project of the nation-state seeks to safeguard it so that it can be used to give the impression of the continuity of the nation. Why then didn't Japan follow the same path? And why did it begin to become part of the World Heritage institutions only some 20 years ago? This chapter will answer these questions.

The rise of imperial ideology

In the previous chapter, we have shown that from the 1850s, the intellectuals had begun to identify the burial mounds. Each burial mound was attributed to an emperor, *Tenno*. This act was directly linked to the rise of the imperial ideology. In the nineteenth century, two schools appeared which aimed to find the foundation of Japan using the imperial family. The first is the Mito school, which belongs to the Mito clan. The school's vocation is to write the history of Japan.

DOI: 10.4324/9781003014904-5

While writing history, it discovered, at the end of the eighteenth century, the difference between Japan and China. In China, there was the reversal of power and consequently, several dynasties. On the contrary, in Japan, there is only the one family that rules. It is the unique genealogy of the imperial family in Japanese history, *bansei ikkei*, which is the uniqueness of Japan. It is the rediscovery of the emperor while the political power is in the hands of the shogun. A shogun sits atop the samurai.

The Mito school proposes the notion of *kokutai*, literally the body of the nation. The *kokutai* represents the unity of the nation, which is a sign of stability and peace. In this sense, *kokutai* is what should be a social form, an ideal form of the nation. According to the Mito school, *kokutai* is based on the principle of submission to the ruler and respect for parents, *chû* and *kô*. Submission to the sovereign must here precede respect for parents. According to Seishisai Aizawa, one of the leaders of the school, at the beginning of Japan's history, this social form was respected (Aizawa, 1973:94). But with the arrival of Buddhism and later Christianity, the social form deteriorated. This principle is also opposed to Confucianism, then dominant in the class of samurai, in which the *kô* has the supreme value. To redress the social form, Seishisai proposed restoring political power to the emperor. It also put forward two axes for the project of recovery. The first axis is the increase of wealth. The second is the military reinforcement. This idea consists in linking the domestic order and the political order. The people must respect their superiors just as they respect their parents. The feelings toward the sovereign and toward parents are similar. For Mito school, the emperor is placed in the position of the sovereign.

Another school that places the emperor at the center of the state is *Kokugaku*, studies of classical Japanese literature. *Kokugaku* first developed by giving importance to ancient writings like *Kojiki* and *Nihon Shoki*, the first two writings in Japan; *Kojiki*'s writing dates from 712, Nihon Shoki's from 720. These two writings constitute the amalgamation of myth and history. The turning point was at the beginning of the nineteenth century with the arrival of Atsutane Hirata, *ronin*, masterless samurai. With Atsutane, *kokugaku* began to take on a political stance. The thought of Atsutane is extravagant, but it played an essential role especially during World War II as the prosecutor of the imperial ideology. For Atsutane, the truth only exists in *Kojiki* and *Nihonshoki*. Claiming to be based on these writings, Atsutane asserts that Japan must be the center of the world. If this is so, it is because Japan is the first country founded by the founding deities; human history therefore begins with Japan and important cultural products and even scientific discoveries were first born in Japan. The emperor is a direct descendant of the founding deities and, as such, he must one day rule the whole world. All of these statements are obviously wrong. But he began to exert influence on the milieu of merchants and rich peasants. From the beginning of the eighteenth century, the market economy developed in big cities, and it gradually spread all over Japan during the nineteenth century.

Kokugaku places more emphasis on the social policy. Nobuhiro Satô, supported by the thought of Atsutane, advanced the project of the new government which would be realized in the government of Meiji. According to Nobuhiro Satô's argument, the government must ensure that the people do not go hungry, and if there are those who are starving, it must help them at all costs. Poverty is therefore considered the first social problem to be solved. And if the government succeeds in solving it, the people naturally recognize the government's policy. They therefore respect and obey the government. In Nobuhiro's project, the emperor is at the top of the newly founded state, so he will be recognized as the nation's greatest benefactor (Satô, 1977:526–527). Nobuhiro already constituted the project for the colonization of East Asia in 1823, the ultimate objective of which is that of China. It is a very detailed project. The project recommends three routes to achieve this goal. The first route passes through the north and colonizes Manchuria and Mongolia. The second route leads to Beijing by conquering Korea. Finally, the third route passes through Taiwan to reach Zhejiang province. When the army of the emperor conquers this entire region, the new capital will be established in Nanjing. This is almost exactly what will happen in the future. *Kokugaku* and the thought of Atsutane caught the attention of scholars of the Mito school. Both schools helped to increase the importance of the emperor. He was attributed to the role of the sovereign for a new government that could replace the shogunate in order to end the feudal system. From the 1850s, this ideology has spread to different social classes.

Mito school and *Kokugaku* put forward the same type of argument. They consider that there are two types of danger. The first danger comes from the rapprochement of Western countries to the archipelago. This is a threat to Japan, and measures must be taken against this threat. The second danger is the uncontrolled development of the market economy. It makes the rich get richer and the poor poorer. Economic inequality is thus accentuated. Poverty endangers the established social order. Both schools agree on the need for a new state in which the emperor reigns to overcome the crisis in society. To this end, the imperial ideology elaborates the project of the imperial system which extends the privileged relationship of the villagers to their ancestors to the relationship between the villagers and the emperor, who is conceived as a direct descendant of the founding gods of the Japanese nation. The people are therefore indebted to the founding imperial family of the nation and must recognize its benevolence forever. Imperial ideology thus allows the state to intervene directly in closed communities based essentially on the order of recollection.

Absence of museological desire

The imperial system was realized in 1868 by the Meiji Restoration. This system was not really interested in constituting the cultural heritage institutions[1] as Western Europe, especially France, which had openly developed museological desire and notably made the aesthetic gaze by placing the Louvre in the center of the republic.[2] In Japan, the museum culture, thus, did not develop in the same way as in Europe. The official history of the Tokyo National museum considers the exposition of Tokyo, which took place in 1872 as the origin of the modern museums. This exhibition was organized to create a national museum later on, in accordance with the decree of May 23 on the conservation of the heritage. In fact, the exposition only gathered objects of curiosity indistinctly. The orna- ment decorating the keep of the Nagoya castle had much success with the public, which took the exhibition for a kind of folklore festival. In an exhibition organ- ized in Nigata, the public brought a small mite to a statue of Buddha exhibited like an ordinary religious practice, since the public didn't regard the statue of Buddha as an object of art.

Tsunetane Sano, which took part in the World Exposition of Vienna in 1873 as a representative of the government thought that the national exposition and the museum, should have the same vocation: the education of the people and the development of industry and the trade. When in 1875, the Ministry of Inte- rior took over the project of the opening of a museum, the then home secre- tary, Toshimitsu Okubo, a very influential character in the government, almost entirely approved the idea of Sano, because the confusion between the world exposition and the museum brought by Sano corresponded well to the policy of the government. It brought closer the economic and the cultural.

According to the suggestion of Sano, national expositions were organized peri- odically starting from 1878. They had as a model the European world exposi- tion, which was different from the museum by nature. When around 1889, the Imperial Household Ministry started the project on the construction of three imperial museums, the orientation of the cultural policy changed more or less. The national museums were, thus, devoted exclusively to the collection of the valuable artistic or historical articles. The national exposition continued to be regularly organized while keeping its ludic and folkloric character. In the fifth national exposition in Osaka in 1903, the merry-go-round and the large eleva- tor were exposed beside the exhibition of the fine arts, and it took a form of an amusement park. The confusion between a museum and an exposition or even an amusement park continued to remain.

The birth of politico-ritual institutions – the case of Ise Jingu

The museum of the European type developed little after the Meiji Restoration, when a different type of cultural heritage conservation was created. This can be called the politico-ritual institutions of heritage preservation. With the institu- tionalization of the imperial system, valuable monuments are attached to the imperial family. They are considered imperial treasures. Everything that is judged to be valuable belongs to the category of imperial treasure.

Even before the Meiji Restoration, the movement for assigning burial mounds to an emperor had begun. The Meiji government tried to accelerate the move- ment. Emperors' burial mounds were not cultural heritage. They belonged to the imperial family. Furthermore, the Ministry of the Interior took care of the objects kept them in Shôsôin, which belonged to Tôdaiji in Nara (Figure 4.1). The shôsô is a repository of the Nara era (710–794), the only one remaining today is that of

Tôdaiji, a temple which has a huge statue of Buddha. This only shôsô is called today Shôsôin. It preserves various objects of the eighth century, for example, offerings at the time of the inauguration of the statue of Buddha in 752. These treasures of Shôsôin remained in disorder until 1892, when the Ministry of the Imperial House took care of their restoration (Tokyo Kokuritsu Hakubutukan, 1973:380). This is in the context of giving more importance to Nara as one of the ancient capitals of Japan. At the end of the Edo period, the burial mound of the first Emperor, Jinmu, was identified with Nara. There is no historical evidence to prove the existence of this person. His name can only be found in *Kojiki*. He is therefore only a mythical character. But, this is how Nara, as the ancient capital of Japan, was discovered or better invented. The historian Hiroshi Takagi points out that until this radical transformation in the history of the imperial family, the beginning of the imperial family was the establishment of the imperial palace in Kyoto in 794 and the fact that there was the imperial palace in Nara between 710 and 794 was ignored (Takagi, 2002:251).

Ise Shrines (Figure 4.2), Ise Jingu, the most important place of pilgrimage in Shinto, also underwent the great transformation. Ise Shrines best demonstrate the integration of religious establishments into the imperial system. Ise Shrines are dedicated to Amaterasu Ômikami, who is considered as Founder of the Japanese nation in *Kojiki*. And the emperor is considered a direct descendant of the founder of the

Figure 4.1 Tôdaiji
Source: Photo by the author.

nation. But John Breen's study shows that, in fact, Ise underwent a significant transformation after 1868 (Breen, 2015). From then on, Ise Shrines became subject to state control. In 1869, Emperor Meiji visited Ise. He was the first emperor to visit the Shrines. From this event, the relationship between Ise Shrines and the imperial family changed radically. At the level of rituals, there was "the merging of ritual cycles of imperial court and Ise Shrines". "The inner Shrine kami hall is now wrapped fourfold in wooden fences, so that spaces easily accessed" before, "are now sealed off" (Teeuwen and Breen, 2017:169). The space of Ise Shrines was rearranged so as to adapt them to the purpose that the nation-state had set. The most popular place of pilgrimage became an ideological apparatus of the state.

At the same time, Ise Shrines created the prototype of conservation/non-conservation of cultural heritage in Japan. If the term conservation/non-conservation is used here, it is because the divine palace is rebuilt every 20 years. In this practice, the material preservation of the palace does not make sense. From the moment the old palace was demolished, it lost the authenticity defined by World Heritage institutions. The new palace is no longer authentic. But, Ise Shrines have the concept of authenticity completely different from that of World Heritage institutions. Ise Shrines claim that the new palace is authentic because it still maintains the original style. Authenticity lies in repetition, in the production of the same. The fact that the palace is demolished and rebuilt every 20 years precisely ensures authenticity. On the contrary, in the practice of

Figure 4.2 Naikû (Inner Shrine) of Ise Shrine. Pictures can only be taken outside
Source: Photo by the author.

Ise Shrines, everything that is material deteriorates and has to be periodically renovated. New is considered pure. What is no longer new must therefore be undone. It is not a simple reconstruction from an architectural point of view. The construction accompanies a whole series of rituals and ceremonies. It is therefore itself a religious practice. Once they are prepared, the Holy Mirror, a symbol of Amaterasu-Ōmikami, is moved to the new sanctuary by the priests.

There are two other characteristics of Ise Shrines. The first characteristic is the presence of the apparently empty space. Ise Shrines always keep an empty space next to the sacred palaces reserved for the reconstruction of future palaces (Figure 4.3). There is also another type of empty space reserved for ritual practices (Figure 4.4). Nothingness has, in fact, an important, even decisive meaning. The second characteristic is the creation of secret spaces. As Breen remarked, pilgrims cannot see all the sacred palaces today because of the redesign of the space made by the Meiji government. Several "living national treasures" that will be explained later in this chapter participate to make the sacred apparel, furnishings, and divine treasures at the time of the reconstruction of the palaces. But Ise Shrines never reveal their names to the public.

Other religious movements born in the Edo period joined the imperial system. Such was the case of the belief in Mount Fuji. The Fujisan as a World Heritage site is composed of two parts. One is "source of artistic inspiration". Another is Fujisan as "sacred place". If Fujisan is a sacred place, this is because it was a high place of pilgrimage. "There are two types of pilgrims, those who were led by

Figure 4.3 Gekū (outer Shrine) and *kodenchi*, empty place reserved for the reconstruction

Source: Photo by the author.

mountain ascetics, and from the seventeenth century onwards, those in greater numbers who belonged to Fuji-ko societies that flourished in the prosperous and stable Edo period" (UNESCO, 2020k). For example, the lodging houses for pilgrims called "*Oshi*" are a component part of the site (Figure 4.5).

According to the doctrine of Fuji-kô, Mount Fuji is considered the source of everything in the universe and, as such, it gives order to the world. Mount Fuji is an absolute benefactor and everything that exists in the universe owes gratitude. On the other hand, Fuji-kô announces the arrival of a new era. And the believers actively spread their belief, and thanks to this spreading activity, the believers increase rapidly. This is why the shogunate strictly controls the activities of Fuji-kô from 1742 (Yasumaru, 1971). But in spite of this pressure from the shogunate, the number of believers continued to increase (Miyata, 1989:144–145).

Most of Fuji-kô's representatives come from the merchant class. This means that the nascent bourgeois class needed a belief and an ethic that could satisfy them. And just like the imperial ideology, the doctrine of Fuji-kô insisted on the notion of *kô*, the debt and gratitude to that which founded life and provided security. The sect did not yet suggest submission to the emperor, because what had to be respected was the sacred Mount Fuji. That said, in the doctrine of Fuji-kô, there was an expectation of transcendental existence. It attributed this role to Mount Fuji. And this disposition to transcendental existence prepared and facilitated the reception of another candidate for transcendental existence, the

Figure 4.4 Empty place for ritual practices in Ise Shrines
Source: Photo by the author.

Figure 4.5 A House of *Oshi*, house for pilgrims
Source: Photo by the author.

emperor.[3] Moreover, Fuji-kô did not deny the existence of the emperor. Fuji-kô borrowed from *Kojiki* and *Nihon shoki* a part of their doctrine. A representative of Fuji-kô went to see the emperor to obtain official permission from their religious society. Some branches of Fuji-kô joined *Kokugaku* and Shinto.

Other systems of cultural transmission

Independently of the political and ritual institutions of heritage preservation, there were various cultural activities. Since the Meiji government was not really interested in cultural activities that did not serve to promote the imperial system, those who led cultural activities had to adapt to the social change produced by the Meiji Restoration. Two systems then emerged.

The first system has their own institution to keep their tradition. This is *iemoto* system in which descendants of the founder, *iemoto*, monopolize the right to deliver the diploma to their followers. This is the case for flower arrangement and tea ceremony. After the Meiji Restoration, the samurai lost their power, and the families of the founders of the tea ceremony and flower arrangement lost their supporters. So they came up with the idea of developing a proper learning system.

They delivered the diploma to those who wanted to become professors. Certificate professors founded their own school in order to teach followers who could become professors in their turn. Many girls' schools introduced tea ceremony and flower arrangement in the learning program. Tea ceremony is especially composed of multiple interactions between the host and the guest. All participants of the ceremony have to respect the implicit, delicate rules of the ceremony. Followers learn from their professor manners, what they should do, and what they should say. If *iemoto* chooses this system, this is because the tea ceremony is held only among intimate friends and not opened to the public. The ceremony is essentially interactional. The *iemoto* does not need to resort to public representation. The second system is the player–audience system. Players of Noh, Kabuki, and Joruri (Japanese puppet theater) play at the theater in front of the audience. They need a larger public that is interested in their art and performance. In particular, Kabuki had its theaters and gathered its spectators since the Edo period.

The worsening situation after the defeat

After the defeat of the Pacific War, large cities such as Tokyo and Osaka were almost in ruins. At the same time, politico-ritual institutions of heritage preservation came to an end. Interest in cultural heritage preservation temporarily appeared under the leadership of GHQ Supreme Commander Douglas MacArthur and repair work was carried out on Kyoto's abandoned and neglected Nijo-jo Castle, today a component of the World Heritage site, the Historic Monuments of Ancient Kyoto. But on the whole, however, the society did not have the capacity to take care of the cultural heritage in the post-war ruins. Those who worked in cultural activities were also struggling. Noh was on the verge of annihilation. Noh masters no longer had any patrons to support them; Noh had a limited audience and was reserved for the bourgeois class that supported it. Some were having trouble making a living (Takahashi et al., 1955). It was the same for some other traditional arts.

Nevertheless, the situation improved from 1950 onwards. First, with the legislation of the Religious Corporations Act, Ise Temple and Tôdai-ji became religious cooperations. Thus, they were no longer directly linked to the state, and they gained the freedom to look after their property. On the other hand, the Act on Protection of Cultural Properties was enacted in 1950. The Law advanced the notion of important intangible cultural property that would establish a unique system in Japan. The act covers cultural fields as vast as theater and ceramics. The exceptions are tea ceremony and flower arrangement. They do not need to make use of the important intangible cultural properties system because they have the *iemoto* system, which is based on the relationship between *iemoto*, founder's family, professors, and followers.

The notion of intangible cultural property focuses on skills for the purpose of protecting traditional performing arts like Noh from the threat of extinction, since preserving only implements like the Noh mask does not help to improve the situation of the performers. Seiichiro Takahashi, Executive Director of the Tokyo

National Museum, argued that there were a lot of Noh-related implements that were not in and of themselves worthy of being displayed at the museum. He noted that the museum could neither collect all of the various dramatic implements nor help promote the revitalization of Noh. Important intangible cultural property refers to the preservation of technical models or skills that are intangible. According to the Act, from 1954, some artists were designated as possessors of important intangible cultural property. They are also called *Living National Treasure*.[4] Living National Treasure is originally a journalistic term which spreads more than the legal term. These two terms are often used interchangeably, but they have rather different meanings. The phrase used in the law does not refer to "living" people, but to the preservation of technical models or skills that are intangible. The journalistic term Living National Treasures, by contrast, refers specifically to artists. Players and journalism prefer the latter one, as it is easier for the public to comprehend.

This new system provoked controversies. Especially critical of the concept of intangible cultural property was Muneyoshi Yanagi, art critic and promoter of traditional art, who felt that cultural properties in pottery had to be tangible. Yanagi felt that value should be judged based on objects with material form. He expressed misgivings about judging handcrafts merely from the perspective of the techniques used and engaged in vigorous debates with members of the Cultural Properties Protection Committee, especially with Fujio Koyama (Yanagi, Koyama et al., 1955). This suggests that the original intangible cultural property system, which was designed based on performing arts like Noh, contained some elements that were not very well suited to arts like pottery, Yanagi's specialty. Yanagi criticized the intangible cultural properties system for its lack of clear selection standards.

The logic of actualization

In spite of Yanagi's criticisms, the intangible cultural properties system is supported by a single consistent logic which we will refer to here as "the logic of actualization" (Ogino, 1995). To speak of a living national treasure is to support the idea that tradition does not always reside in the works already realized, but in those that are in the process of being realized. Traditional art does not consist, therefore, in faithfully preserving the heritage of the past, but in actualizing in the present what once existed or is believed to have existed in the past. According to this perception, tradition as such does not exist or, even if it does, it is invisible and must therefore be made manifest in order to be recognized. This logic of conceiving tradition is precisely the logic of actualization.[5]

Traditional theater, like Noh and Kabuki, shows, better than other forms of traditional art, this conception of immateriality, because the actors' performance only manifests itself on the stage. Dramatic performances are first given "form" when they are performed by actors and performers. Without flesh-and-blood performers, the transmission of the tradition would be impossible. Thus, when an actor, designated as a living national treasure, performs in the national theater, the spectators witness the tradition that his individual performance makes manifest. Here tradition has no fixed figure, and therefore remains invisible until it is staged. And it is precisely the living national treasures that actualize it. The theatrical performance thus takes on the form of a heritage exhibition performed by recognized actors and musicians. In this sense, tradition is less about faithfully transmitting a legacy from the past than about understanding a particular kind of performance as tradition in the present. This is different from the European approach of passing down a fixed tradition from the past into the present and then on into the future.

In the logic of actualization, the body plays an essential role. The protection of intangible cultural properties is based on the body. Actualization of what is supposed to have existed in the past is possible through the body of owner of intangible cultural property. Tacit skill and knowledge belong only to some excellent individuality. It is realized through their performance. In many cases, these individuals are trained within a hereditary system like Kabuki theater. It seems a proof of their traditional character.

Principle of secrecy

In the logic of actualization, those who create the works count above all, and the conservation of their works is secondary. Ise Shrines illustrate this well. They are not interested in preserving forever materially the same sacred palaces. On the contrary, periodic reproduction repeats itself ad infinitum. This practice resuscitates what is original. Renewal is, in this conception, the recreation of the original.

Nevertheless, there are objects and buildings that have withstood the wear and tear of time. There are many antiques and collections, including pieces of art that have been designated as National Treasures, *kokuho*. This concept was legally defined already in 1897. The Law for the Protection of Cultural Properties keeps the concept. What is then the attitude with regard to these objects and monuments?

At the Louvre, history is presented to us by the objects preserved. They are exhibited to the public, and they are always visible to the public. This visibility of preserved objects is essentially ensured by the museum. On the contrary, the objects are kept away from the public in the logic of actualization. In fact, there is no need to ensure the presence of linear history through the objects, and even if there is the will to preserve them, the public does not see it clearly. This characteristic of keeping distance with the public can be defined as the principle of secrecy. The example of Ise Shrines is a good illustration of this lack of openness. This is the same for Shôsôin. The treasures are transported today to the newly built depots in order to preserve them better. The public is only allowed to have access to a part of them, and this only during the annual exhibition which takes place in the Nara National Museum. The place of conservation and the place of exhibition are therefore distinct. The exhibition of the treasures resembles an actor's entrance on stage. The treasures are invisible in ordinary times, and the secret of the treasures is only revealed at the time of their exhibition. This invisibility gives the objects a mysterious character. The public discovers extremely rare, even exotic things rather than historical objects passed down from generation to generation. The objects are not there to represent the past, they help to show that they are still in the present, though simply hidden in ordinary times. Other temples and shrines that own national treasures only make them available to the public once a year for a limited period of time or establish their own museums, refusing to actively display national treasures in national and other public museums.

The same logic applies to many historic buildings. For example, in Kyoto, palaces and a good number of temples only open on certain days. This periodic opening lets the public know that these monuments are still in use and even inhabited today. In this way, the public discovers a world spatially removed from everyday life that updates tradition. This kind of secrecy is not altogether a bad thing from the perspective of preserving cultural heritage, since objects tend to deteriorate the longer they are exposed. Japan has a culture of storage that is based on the premise that refusing to display objects publicly helps keep them preserved for a longer period of time. This is explicitly the principle adopted by the Shôsôin.

Public and private

In the politico-ritual institutions of heritage conservation, all that could be used for the promotion of the imperial system became the object of state intervention. But the government was not really interested in everything else and left quite a large margin of freedom. The protection of cultural heritage was, thus, largely relegated to the discretion of owners. As a result, even items discovered to have great value as cultural heritage objects were kept under the ownership of individuals, not purchased by public institutions. Making them available to the public or donating them to museums was still uncommon. The world of antiques was based on a principle of secrecy in which people only wanted to share their high-value pieces used for formal tea ceremonies with a limited group of close friends. Mounting work was done for prominent people who were able to hold tea ceremonies.[6]

The relative autonomy of the private sphere and the passive attitude of public authorities toward the acquisition of cultural heritage continue to exist after the disappearance of politico-ritual institutions. In one town especially renowned for its pottery an enormous kiln used in the Edo period was owned by a ceramicist who did not have a good relationship with the local authority. For this reason, the kiln which should have been designated as a cultural property was not, and the local government did nothing to step in. Today, the owner of the kiln is deceased. Then, the museum he created is definitively closed. The kiln is not yet classified as cultural heritage. Another example is that of a tree. If the roots of a tree that is eligible for designation as a natural monument are growing on private land, those who wish to designate the tree as a monument may encounter opposition from the landowner.

By way of comparison, we will give the example of France, which contrasts with the Japanese case with regard to the public–private relationship. In France, the national government would actively strive to bring such a piece of land under state ownership. For example, the Conservatoire du Littoral, founded in 1975, is promoting the buyback of coastal land that became private property during the nineteenth century. The purpose of this effort is "to restore the public's rights" to that land according to an expert of the Conservatoire. Specifically, the work of the Conservatoire du Littoral is to buy back coastal lands that were monopolized by the bourgeoisie and open them up as public spaces. In this case, the coastal lands are clearly viewed as public property. In addition, they are also viewed as a type of cultural property. This is difficult for the Japanese to understand, given their ideas about nature. To the Japanese, the coastline is a given, and it is difficult to see it as something that is consciously owned or that should be protected by the state as a cultural property.

In France, the state is in charge of publicness, so if something is deemed to have a certain degree of publicness, the state can actively intervene in the private sphere. In this kind of situation, publicness is achieved in two ways. First, the state strives to fulfill a demand of citizens deemed to be of public value. The opening up of the coastline to the citizens is the case. Second, the state acts as a coordinator in handling the conflicts and tensions that arise. The role of the coordinator is widely varied and may range from achieving a balance between environmental protection and tourism to figuring out how to handle discord between local hunters and the EU environmental standards for protection of birds. The Japanese administration lacks this kind of coordinating role and takes a passive approach to responding to requests by citizens. In Japan, the government does not tolerate interference by the private sector in areas under its jurisdiction and tends to consider that its decisions are beyond reproach. The government can operate at its own discretion. But the public authority cannot easily interfere with private property. The truth is there is a kind of segregation between the public and the private spheres, with rigid boundaries that divide the two (Ogino, 2002: 225–226).[7]

Nevertheless, it should be noted that in France also, families have a way of preserving their legacy. Pierre Bourdieu studied how farmers in the southwest of France tried to conserve their familial heritage. In the traditional French community, farmers established a system that allowed them to preserve their legacy by giving all property to an only child, in general the eldest son (Bourdieu, 2002). Houses, furniture, and other property constitute familial heritage, and all these objects and constructions are transmitted from generation to generation. The property of great farmer families has not only a material meaning, but it acquires a symbolic sense. Families have, thus, an affective relationship with their legacy. The owner of a large rural property of a French family in the south of France continued to keep not only his house, garden, and fields but ancient equipment for viticulture in the shed. It can be regarded almost as an eco-museum. It is just only "almost", because the shed is not open to the public. Families and the nation have their own way of preserving heritage in France just as in Japan. Only, in Japan the public power intervenes less directly in the preservation of cultural heritage which belongs to individual properties.

Bizen ware

While the private sphere is relatively autonomous, the legislature on the protection of cultural heritage in 1950, and, in particular, the concept of the intangible cultural property served generally well for the recovery of villages and towns that were having a holder of intangible cultural property. To show this, we will take the case of Bizen. Bizen is a city of pottery that dates to the twelfth century. The characteristic of Bizen ware is that it never uses varnish. It is thus famous for its natural beauty, and this is the reason why it is used in the tea ceremony, which values the beauty of the dark, *wabi*. Bizen thrives today with 123 art ceramicists who run their store next to their pottery kiln.

After the legislation, the first holders of intangible cultural property are registered, including four ceramicists. Muneyoshi Yanagi divided these ceramicists into artists like Shoji Hamada and Kenkichi Tomimoto, whom Yanagi himself had supported, and other ceramicists like Munemaro Ishiguro and Toyozô Arakawa, whom Yanagi criticized as being unworthy of recognition because of their lack of creativity. In spite of Yanagi's underestimation, Ishiguro is renowned for the rediscovery of the ceramic technique that existed in the Chinese Sung Dynasty[8] and Arakawa for the revitalization of Shino and Setoguro (black Seto) style porcelain in the Momoyama-period at the end of the sixteenth century in Japan. Arakawa even discovered the ruins of a kiln on his own and used some fragments of pottery found near that site to revive the Shino style of porcelain. These are typical examples of how traditions can be actualized. The pieces Arakawa actually produced were new creations, but he made them using traditional techniques.

Among the first four registered, there were no ceramicist from Bizen. But, the following year, Tôyô Kanesige was designated as holder of intangible cultural property, living national treasure. Kanesige came from an ancient family of potters and revived the tradition of Bizen that had been for a long time in a state of stagnation. The crisis of Bizen ware was overcome thanks to the strong personality of Kaneshige. After Kaneshige's death in 1967, there is no longer a living national treasure in Bizen. This means in a way the disappearance of a part of heritage. For the next three years, there was no ceramicists worthy of the title of living national treasure, and it was expected that the same art value as Kaneshige's would be resurrected by someone else. This was realized in 1970 with the designation of Kei Fujiwara. Tôshû Yamamoto became a living national treasure in his turn. Each time, the nomination of a living national treasure is experienced as the event of heritage transmission. The history of Bizen ware is thus represented by the living national treasures which are considered as embodiments of the traditional art of Bizen.

At the request of the Bizen ceramicists, Okayama Prefecture, to which the city of Bizen belongs, has been organizing courses since 1966 and established the Bizen Ceramic Research and Training Center in 1971. Thanks in part to this center, the number of art ceramicists has increased. Trainees, all uninitiated, spend a year there to learn the basics of the ceramic art of Bizen, whereas it would take them at least three years if they worked as apprentices with a ceramicist.

But this opening to the outside world – some trainees come from far away – does not modify the structure of Bizen ware; ceramicists and all the people who gravitate around them organize their milieu by taking advantage of the legal framework; and the essence of this structure is to present this milieu above all as a world of artists that possesses at its summit a living national treasure. On the contrary, the center contributes to reproducing the structure by enriching it. To observe this process of reproduction, it is enough to look at the life course of the trainees at the center. After one year of internship, four possibilities are offered to them; to work in a factory, to be a disciple of a master, to return to the family if they are ceramicists, or to establish themselves immediately on their own; the last possibility is nevertheless extremely limited. Whichever possibility they choose, they will be classified into two categories: either that of an artist or that of a simple potter. In the future history of Bizen ware, only those who will be part of the category of artists will be taken into account. Apart from them, all those who will have worked for Bizen ware will be eliminated from the history, or at least left aside.

Economy and cultural heritage

The existence of the living national treasure helps to activate the ceramic market. Indeed, the designation of a ceramicist as owner of the intangible cultural property can increase the rating of the place where he/she does his/her activities. Indeed, as Figure 4.6 shows, the market value of the works of ceramicists who have been designated as Important Intangible Cultural Properties is rapidly increasing. The appraised value of works by three such designated artists, Kyusetsu Miwa, Uichi Shimizu, and Tôshû Yamamoto, has steadily risen since their designation. The appraised value of Shimizu's works rose especially rapidly after 1985, when he was designated, and have surpassed the value of Yamamoto's works, who was designated two years later. Works by Miwa, who was designated the earliest of the three in 1983, were the most highly appraised among all ceramicists in Japan in 1994. The appraised value of each of these three artists' works at any point in time correlates with their designation as an important intangible cultural property. Ceramicists make use of Important Intangible Cultural Properties system and a journalistic and more popular term, living national treasure, because they also need a public that appreciates their products like Kabuki and Noh players; their activities depend largely on the market. Local communities and artists themselves do not hesitate to take advantage of the market to conserve heritage, and they seek to continuously link the economy and conservation strategy. A researcher at the Bizen Research and Training Centre said in the interview, "we will benefit greatly from the shinkansen line especially for the first ten years. It is always necessary to readjust to the situation. Otherwise, the traditional art declines".

This contrasts sharply with the cultural policy of France and Western Europe. In France, an object can only acquire real historical value when it is exhibited in a museum. It has to be permanently out of the market to be among the important historical objects. Jean Guibal remarked, "I am always very surprised by the fact that museums only give value to objects when they have acquired a relative

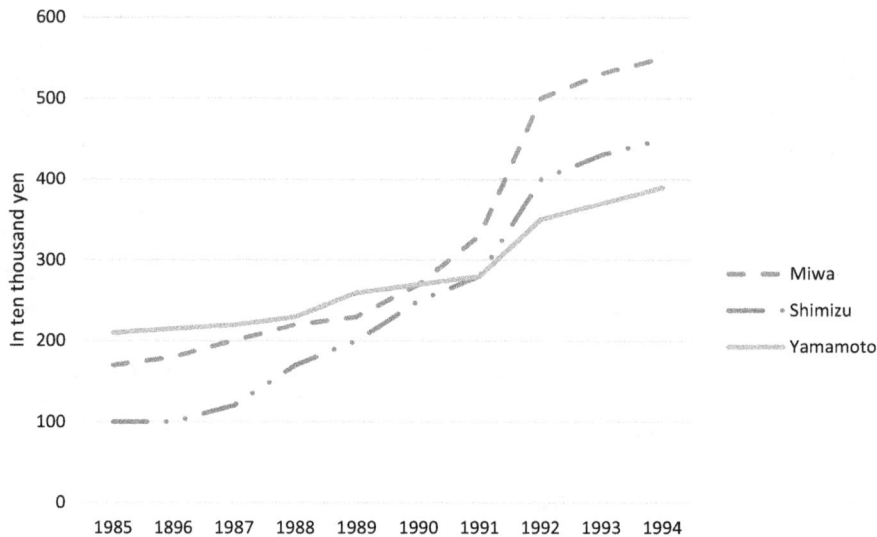

Figure 4.6 Evolution of the rating of the three artists (1985–1994) (Ogino, 2002:214)

notoriety at collectors. It will therefore be necessary for the Museum of Ethnology to resume its collection work" (Guibal, 1992:160–161). The dependence of museums on the market for the selection of objects is an aberration, and the museum must always lead the construction of heritage. The museum is therefore not simply a place of exhibition, but the agency prevents the market economy from intervening in the constitution of heritage. This separation of economy and heritage is indispensable to ensure the linear representation of history; all the more so since the market economy hinders the will to conserve by creating the fluidity of objects. In order to satisfy museological desire, the museum must detach the objects from their environment in which they are used. This uprooting of the objects is still not desirable for the Bizen ware circles, who hope above all to sell their products and who, consequently, attach little importance to putting great works in a museum.

The example of Bizen shows how institutionalization of intangible cultural property contributes to sustain the production of pottery. So, do all cultural activities work like Bizen? The answer is no. The constitution of the Important Intangible Cultural Properties system did not always result in the quick revitalization of all arts. This is the case for puppet theater in Awaji Island. Once there were many puppet theaters. But the number of theaters has been gradually reduced, and before the Ohnaruto Bridge was built in 1985, only one theater gave minor performances on the second floor of the waiting room of the 24-hour passenger ferries that ran between Awaji Island and Naruto. Then, Tomoji Tsuruzawa was designated as an important intangible cultural property in 1986 in

Figure 4.7 Awaji Ningyô Za (Awaji Puppet Theatre)
Source: Photo by the author.

the field of *shamisen*, a Japanese guitar that accompanies the puppet play. Puppeteers gave periodic performances at the Awaji Ningyo Joruri Kan. However, with inadequate financial support, the conditions they face made it difficult to put on satisfactory public performances. The situation has slowly improved. Puppet theater is taught in elementary and junior high schools on Awaji Island; young people with an interest in the art can become members of the theater after they graduate. Thus, a new generation of puppet masters is being cultivated. In 2012, a new theater (Figure 4.7) was finally constructed.

Cultural transformation from the 1990s

The politico-ritual institutions and the two systems that we called *iemoto* system and player–audience system, which appeared after the Meiji Restoration, have the same characteristics. First, they do not give the primary value to the material conservation of objects and monuments. Second, they give respect to empty space and nothingness. Third, the greater the importance, the more secrecy must be kept. Transmission of skills also keeps secrecy. And finally, they are based on the notion of non-linear time. The intangible cultural property system was born at a time when efforts were being made to maintain these characteristics within a modern institutional framework. The Law for the Protection of Cultural Property can be considered a result of the adaptation of the politico-ritual institutions and player–audience system that were no longer legitimate in the post-war regime by objectivizing, through legal clauses, the underlying logic of actualization.

But the situation began to change around the 1990s. Showa Emperor Hirohito died and the new era, Heisei, begun in 1989. The situation changed with regard to the issue of cultural heritage with the death of the emperor, who experienced both the pre-war imperial system and the status of the emperor as a symbol in the post-war democratic regime. The parliament started to discuss the ratification of the World Heritage Convention in 1988 and finally ratified it in 1992, 20 years after its adoption. Japan was the 125th country to conclude the convention. At the same time, museological desire began to develop little by little throughout Japan. Local governments competed with one another to build museums. At the beginning, because they were unable to collect any vital elements of cultural heritage, they often left the inside of these extravagant buildings bereft of any real content and were forced to buy modern Western paintings at exorbitant prices. That said, some of the local governments and communities tried to discover their cultural heritage and some of them have made efforts to have their heritage inscribed onto the World Heritage List. The idea that cultural heritage is something public is, thus, beginning to be shared. In consequence, the principle of secrecy loses its importance little by little.

On the other hand, political and social movements have helped to spread the idea that cultural heritage should be public. The pacifist and anti-nuclear movements have encouraged the creation of peace museums. This led to the inscription of the Genbaku Dome on the World Heritage List. The ecologist movement followed the same path. One example of the ecologist movement, which gave birth to the constitution of cultural heritage, is the movement promoting the designation of the Ashio Copper Mine as a World Heritage site. The nomination file for World Heritage Tentative List was submitted in 2008. The title is Ashio Copper Mine: The Starting Point of Japan's Modernization, Industrialization, and Pollution Control. It summarizes the contents of the site as follows:

> The period of 100 years from the late 19th century was an era in which pollution became a social problem in the industrialized world. The Ashio Copper Mine and its surrounding area, which was devastated by rapid development and smoke pollution, and the monumental buildings that illustrate the history of pollution control technology, are outstanding examples of the technology and landscape that marked this period. It is a living example of what we should be doing today.
>
> (Tochigiken Nikkôshi, 2020)

The Ashio Copper Mine, exploited by Furukawa zaibatsu after Meiji era, is known for the first case of serious pollution in Japan. Indeed, the copper mine polluted the Watarase River, which in turn caused devastating damage to the villages along the river. To put an end to the pollution, in 1906, the Meiji government decided to make a reservoir to absorb the polluted water in a village, Yanakamura, without blaming the Furukawa zaibatsu, which operated the mine.

The villagers were thus forced to leave their village. Some villagers resisted but to no avail.

Much less known than the case of Yanakamura, there was the disappearance of another village. This was Matsukimura, Matsuki village. This village was located not far from the Ashio Copper Refinery (Figure 4.8) managed by Furukawa zaibatsu from 1878. Matsukimura was largely affected by the smoke pollution but Furukawa hid what had happened. Thus, villagers could not lead a normal life due to the excess of smoke pollution, and they were forced to leave definitely the village in 1901 because of pollution. Since then, all traces of this village have been erased. Today there is no one left from those regions (Kamioka, 1987).

The movement to make Matsukimura a cultural heritage organizes a type of pilgrimage to the site where the village once stood, where few gravestones still remain. Those who will participate in this movement by walking to the Matsukimura ruins do not seem to have the clear intent of going on a formal grave visit. Nonetheless, the actions of those who are looking for meaning in the ruins of this village have gone beyond ecological movement and are taking on the characteristics of a pilgrimage. Promoters of the movement firmly believe that by reconsidering history, they can be spokespeople of the farmers who had no choice but to leave Matsukimura because of the pollution. Relating a history that has been forgotten seems to make it possible to resurrect the voices of those who

Figure 4.8 Ashio Copper Refinery out of use

Source: Photo by the author.

died from the pollution. This is why the movement has basically turned into a pilgrimage for memorializing those who perished.

This example shows well the transformation that took place in the 1990s. On the one hand, the new conception of cultural heritage is spreading, and it legitimizes the idea that cultural heritage should be shared by the public. But the logic of actualization is also present in the case of Matsukimura insofar as this movement consists in recalling what had been forgotten. Here appears the importance of the narrative considered as collective memory. Even if well-preserved objects and monuments are absent, it does not matter. The constitution of a narrative considered well-founded is sufficient for there to be a cultural heritage. It makes it possible to create a heritage site even if almost nothing remains in place. It tends to lead people to see the potential for preserving heritage in even the smallest sample of old ruins where no such heritage, or very minimal components of heritage, exist. On the whole, the "modern" institutionalization of cultural heritage in the 1990s still accompanies the logic of actualization. It is still the underlying logic when selecting a heritage.

At the same time, two of three representatives of the ancient politico-ritual institutions, the burial mounds and Tôdai-ji, became World Heritage sites. It means that they adapt to the World Heritage institutions and accept its norms. Tôdai-ji, of which Shôsôin is a part is a component part of the World Heritage site, Historic Monuments of Ancient Nara inscribed on the World Heritage List in 1998. The Tôdai-ji Museum was established in 2011. Forty-nine burial mounds are part of the World Heritage site, the Mozu-Furuichi Kofun Group inscribed in 2019, as shown in detail in the previous chapter. The Sakai Museum is organizing the permanent exhibition which places Daisen Kohun, the largest burial mound, in the center of history of Sakai city. On the other hand, Ise Shrines newly opened a Sengukan Museum dedicated to the ritual of Shikinen Sengu in 2012. In addition, Ise Shrines decided, for the first time in its history, to preserve the ancient ornaments whereas they had traditionally destroyed them all. Some of the ornaments are on display in Ise Shrines' main museum, Chôkokan.

An amalgam of the logic of actualization that suggests the principle of secrecy and museological desire that emphasizes the exhibition of cultural heritage emerges from this process. This is the current situation of cultural heritage conservation in Japan.

Notes

1 Tetsujirô Inoue (Inoue, 1908:474–476), a philosopher who tried to justify the imperial system clearly noted that the notion of *chû* is almost identical to that of *kô*, that is, the people must respect the emperor as their parent.
2 Ukiyo-e contributed to the formation of aesthetic gaze in Japan. But it was considered as popular art, and the nascent aesthetic gaze could not play a role in the constitution of the nation-state as it did in France.
3 Two stories that are both part of the Fujisan site have in common to consider Mount Fuji as unique and transcendental. This expectation of transcendence has

been represented artistically by the ukiyo-e artists on one side and morally and religiously by Fuji-kô on the other. The value of the latter outweighed the former.

4 In 1994, France established a *maitre d'art* system modeled after Japan's important intangible cultural properties system. However, it differs from the Japanese system insofar as it also designates, for example, an individual who has the skills necessary for repairing historical building as a *maitre d'art*. It is close to UNESCO's intangible heritage. Indeed, in 2020, traditional Japanese carpenters are designated as intangible heritage.

5 *Iemoto* system also follows the logic of actualization.

6 Conservation Center for Cultural Properties opened only in 1980 at the Kyoto National Museum. It performs repairs using traditional techniques.

7 The argument that has been strongly emphasized in the social sciences in Japan since World War II, that Japan's state or public sector is extremely powerful while the rights of the private sector are relatively weak, is an unconvincing argument.

8 The Sung Dynasty began in 960 and ended in 1279.

5 Age of preservation

The way of preserving cultural heritage varies from one society to another, as the previous chapter has shown with the example of Japan. However, the World Heritage Convention carries on the standardization of cultural value. Marc Askew noted, quoting Logan's comments, that UNESCO's global heritage system had imposed forms of "cultural homogeneous globalization" onto the non-Western world. UNESCO improvises international practice, promotes particular sets of heritage values and conservation practices, and establishes common management practices in World Heritage sites (Askew, 2010:26–27). There must be different attitudes in the face of this imposing globalization. In the case of Japan, it has rather a solid cultural base and has strategies to adapt to the imposing standards of UNESCO. It has even modified them through the constitution of the Nara Document on Authenticity. But, there are societies that cannot cope with imposing cultural globalization.

Nevertheless, World Heritage institutions do not possess absolute power and are affected by the process of more global change. This is the Age of Preservation presented in the Introduction. The Age of Preservation contributes to the removal of existing boundaries, boundary between the past and the present, barriers between the inside and outside of cultural heritage sites, and the distinction between those who give cultural value to objects and monuments, and others. World Heritage institutions have been altered by the effects of the Age of Preservation. Regions that have a World Heritage site are confronted with the characteristics of the Age of Preservation. There are regions that adapt well to it. There are others where conflicts arise. But whatever the case, societies are more or less forced to change with the arrival of the Age of Preservation.

Transformation of the temporal representation

In cultural heritage institutions originally, monuments, by their very presence, ensure the continuity of history.

The structure of the "cold", that is, non-modern society exists in such a way as to nullify the awareness of the succession of events, and thus of historical consciousness. On the contrary, in the "hot" and modern society, "internalizes

DOI: 10.4324/9781003014904-6

resolutely the historical becoming in order to make it the motor of development" (Lévi-Strauss, 1962:310). Museums, as they developed in the nineteenth
century, were based on the linear conception of time and history just as archives,
according to Lévi-Strauss, the incarnated being of history (op. cit., 321).[1]

However, societies that have voluntarily created cultural heritage institutions
are currently experiencing the effects of the Age of Preservation. We will take
here the example of Lyon to show this change. The historic site of Lyon was
inscribed on the World Heritage List in 1998. Its Outstanding Universal Value is
based on criteria (ii) and (iv). Here is its explanation.

> Criterion (ii): Lyon bears exceptional testimony to the continuity of urban
> settlement over more than two millennia on a site of great commercial
> and strategic significance, where cultural traditions from many parts of
> Europe have come together to create a coherent and vigorous continuing
> community.
>
> Criterion (iv): By virtue of the particular manner in which it has developed
> spatially, Lyon illustrates in an exceptional way the progress and evolution
> of architectural design and town planning over many centuries.
>
> (UNESCO, 2020l)

This explanation emphasizes the historical continuity of the city through
expressions such as "the continuity of urban settlement over more than two
millennia" and "vigorous continuing community". In particular, in criterion
(iv), the city is considered to demonstrate architectural history and urban
planning.

But in reality, the Lyon site corresponds to the Age of Preservation. In the old
town, the red-tiled buildings stretch. The walls of each building are painted in
cream or light pink. These buildings are famous not only for their beauty, but also
for their traboules (Figure 5.1). A traboule is "a narrow passage that consists of
a long passage that spans one or more buildings and a courtyard that leads to it",
and it was used as a way out for the resistance in Lyon, the center of the resistance
during World War II. In the old town, 500 traboules are preserved, and some
of them are open to the public. At the entrances of some of the traboules, there
is a copper plate that says "Mémoire de Lyon" with an explanation of each traboule. However, it is not merely preserved as a memorial. It is still used as a residence and a living space for the citizens (Figure 5.2). There is the quiet daily life
which is not disturbed by tourists-consumers. Fourvière, the remains of a Roman
amphitheater reached by cable car or on foot from the old town, is more touristic
but is not a simple Roman vestige. It is a meeting place for various activities. The
event Les Nuits de Fourvière was held from Tuesday June 7 to Saturday July 30
2011. At that time, sound and lighting equipment was brought in to serve as a
place for musical and theatrical activities. There were also exhibitions of artists'
photos, a circus tent, and a temporary store set up in the area. There is no longer
a barrier between the interior and exterior of the World Heritage site in Lyon.
The site is firmly in the present, and its components are still in use today in one

way or another. Therefore, rather than showing the historical evolution, the city displays its different characteristics spatially.

It is the same for Paris. The site Paris, Banks of the Seine is inscribed on the World Heritage List in 1991. The narrative on the site focuses on the historical evolution of the city and the architectural conservation that clearly shows this evolution. And it can be seen from the banks of the Seine from the Ile St Louis to the Eiffel Tower. "A large number of major monuments of the French capital are built alongside the river and on the perspectives overlooking it" (UNESCO, 2020m). In the explanation of the Paris site, the historical continuity of Paris is emphasized. But at the same time, the originality of Paris lies in the fact that this historical evolution can be observed spatially. What is old coexists with what is new, and this is not a problem. Besides, Parisians live in Paris, and there is really no border between the cultural heritage and the rest.

There are also societies that are not traditionally familiar with cultural heritage institutions but are adapting to the Age of Preservation. This is, for example, the case of Tunisia. Three Tunisian sites were inscribed on the World Heritage List in 1979. The first site is Medina of Tunis (Figure 5.3). The second is the archaeological site of Carthage located near Tunis. The third is the Amphitheater of El Jem. The Medina of Tunis represents a typically Islamic city. There are, in a well-circumscribed space, mosques, the old palace, and the souk, etc. It appears as a labyrinth to strangers to the city. The daily life is carried out in the city and like the old district of Lyon, Medina adapts well to the institutions of the World Heritage. The archaeological site of Carthage (Figure 5.4) is located in the suburbs of Tunis. Near the site, chic residences, including that of the president of the republic, are observed and it is a quiet place. There are relatively few visitors. It is rather a classic World Heritage site and probably "just right" for UNESCO. The Amphithéâtre of El Jem is like the remains of Carthage, a Roman vestige situated in the small village of El Jem. During our survey in 2014, a concert was organized in the amphitheater as part of the International Festival of Symphonic Music of El Jem (Figure 5.5). It was just before the presidential election, and it was the first presidential election in Tunisia. It was just the time of the Arab Spring. Tunisians were more or less excited for this direction started for democratization. The one who composed the music for the concert was a politician, Jaloux Ayed, and among the spectators of the concert, there were some who pointed out the political intention of the event. On the other hand, the exploitation of tourism around the amphithéâtre had just begun. Around the amphithéâtre, there are only a few souvenir stores and restaurants. But it does not seem to be a disorderly development. At least the upper and upper-middle classes recognize the importance of the amphitheater. Answering our question as to whether a Roman vestige could be considered part of Tunisian culture, a student who participated in the concert said that if such an important monument once existed on Tunisian territory, Tunisians should be proud of it. Even if it is of Roman origin, it doesn't matter. Whatever their origin, what matters is their existence on Tunisian territory. Moreover, the expansion of consumer culture seems to be hampered by Islamic morality.

Figure 5.1 Traboules in Lyon
Source: Photo by the author.

Tunisians know by themselves how to manage cultural heritage sites. It is they who create and innovate cultural value by organizing events around World Heritage sites while respecting UNESCO standards. According to UNESCO's estimate about Carthage,

> Restoration and maintenance work carried out over the years is in accordance with the standards of international charters and has not damaged the

Figure 5.2 Mailbox for residents in Traboules
Source: Photo by the author.

> authenticity of the monuments and remains of the site of Carthage. The site benefits from a maintenance protocol.
>
> (UNESCO, 2020n)

The Tunisian case shows that if a site adapts in one way or another to the constituent elements of the Age of Preservation, serious problems do not arise.

Maladjustment to the age of preservation

Unlike Lyon and Tunisian sites, some World Heritage sites illustrate the maladjustment to the Age of Preservation. We will take the example of the Dresden Elbe Valley in Germany. The site was deleted from the World Heritage List due to the decision to build a bridge. It shows the discrepancy between World Heritage institutions and the Age of Preservation.

The construction of a new bridge, the *Waldschlößchenbrücke*, over the Dresden Elbe Valley had been decided in 1996 at the City Council of Dresden, the majority of which was the Christian Democrats (CDU). The Dresden Elbe Valley was included in the World Heritage List in 2004. In the same year, The Social Democrats (SPD), the Green Party, and the Left Party

Figure 5.3 Medina in Tunis
Source: Photo by the author.

Figure 5.4 A poster of classic music concert at Carthage in French language in Tunisia
Source: Photo by the author.

Figure 5.5 El Jem in Tunisia, preparation of a concert
Source: Photo by the author.

won the election and announced that the construction of the bridge would be stopped. Faced with the decision of the new majority, the former majority had the idea of holding a referendum. It was carried out in 2005. About 50.8% of the voters took part in the referendum, 67.92% of whom voted for the construction of the *Waldschlößchenbrücke*. Respecting the result of the referendum, the city council continued the project of building the bridge. However, just after this decision was taken, UNESCO announced the unfavorable opinion on the project of the City Council of Dresden. In 2006, the Dresden Elbe Valley was placed on the List of World Heritage in Danger. The city council decided to stop the construction of the bridge to meet the UNESCO request, but the Mayor of Dresden did not agree and neither did the Free State of Saxony. The city council had, then, recourse to legal proceedings but in vain as the result of the referendum is valid for three years according to the German jurisdiction, and it must be respected. The state party, Germany, and UNESCO did not find a solution, and finally, the World Heritage Committee removed Dresden Elbe Valley from the World Heritage List in 2009 (Figure 5.6).

There was, from the beginning, a misunderstanding between UNESCO and the city council. A researcher of the UNESCO World Heritage Centre pointed out three problems on the part of the city of Dresden in our interview with her in August 2013. First, the inhabitants were not well informed.[2] Second,

Figure 5.6 Aposter shows the removal the Dresden Elbe from the World Heritage List with the sentence "Anger and tears remain"

Source: Photo by the author.

UNESCO was not officially informed of the project, even though the state party was obliged to do so. An expert sent by ICOMOS visited the site just before the submission of the nomination file and was informed of the project for the construction of the bridge. In the nomination file, just few phrases evoked the construction of a new bridge: "The construction of a new bridge is foreseen 5 km^3 down the river from the center. Its design results from an international competition. The profile has been kept slender and low in order to reduce impact on landscape" (UNESCO, 2020o). Third, the state party did not conduct research on the effects of the construction of the *Waldschlößchen-brücke* (Figure 5.7). Thus, UNESCO asked the Technical University of Aachen in Germany to evaluate the effects of the construction of the bridge. The University gave a negative assessment. They gave three following conclusions:

1   The Waldschlößchen-Bridge does not fit in with existing series of Dresden City Bridges;
2   The Waldschlößchen-Bridge obscures a number of views of the Dresden sky-line and the Elbe valley which are of historical importance as well as continuing relevance to daily life in the city;
3   The Waldschlößchen-Bridge cuts into the cohesive landscape of the Elbe River bend at its most sensitive point, splitting it irreversibly into two halves.

Figure 5.7 The *Waldschlößchenbrücke* in Dresden
Source: Photo by the author.

A Japanese specialist questions this assessment, which did not take into account the opinion of the citizens of Dresden (Abe, 2011:7188–7189). Even if the Technical University of Aachen is renowned for this kind of work, the evaluation on the value of the landscape can hardly be "objective". It is difficult to keep the same landscape always in the same state because it is not a still life.

This example seems to illustrate the typical conflict that the inscription of a site on the World Heritage List can pose. Moreover, the problem contains the political conflict of the city. But, from our point of view, this is a typical example that shows the discrepancy between World Heritage institutions and Age of Preservation for three reasons. First, there is discordance between the rigorous application of the standards of protection and the public opinion of citizens. In the Age of Preservation, the latter must be respected whereas in the decision-making of the World Heritage Committee, only the opinion of experts was adopted. Second, the landscape is not something fixed. It is difficult to completely fence off the landscape. In the Age of Preservation, barriers between inside and outside of cultural heritage become blurred. Third, in the Age of Preservation, the boundary between the past and the present is removed. Adding a new element does not necessarily spoil the value of a cultural heritage as we will see later.

Another example of maladjustment to the Age of Preservation is Angkor. About Angkor, UNESCO emphasizes the importance of

> associating the 'intangible culture' to the enhancement of the monuments in order to sensitize the local population to the importance and necessity of its protection and preservation and assist in the development of the site as Angkor is a living heritage site where Khmer people in general, but especially the local population, are known to be particularly conservative with respect to ancestral traditions and where they adhere to a great number of archaic cultural practices that have disappeared elsewhere. The inhabitants venerate the temple deities and organize ceremonies and rituals in their honor, involving prayers, traditional music, and dance.
>
> (UNESCO, 2020p)

There are underlying contradictions in this suggestion. On the one hand, this "management requirement" seems to plead the autogenous development with words like "intangible culture", "living heritage site", or "archaic cultural practice".[4] On the other hand, the Cambodian Government decided on the land use in the Angkor Park in 1994 that all lands located in zones 1 and 2 of the Angkor site were state properties. In order to strengthen and to clarify the ownership and building codes in the protected zones 1 and 2, boundary posts have been put in 2004 and 2009.[5] At the same time, the Royal Government Decree of November 16, 2004 postulates that the citizens who have been residing in those areas for a long time are entitled to abide in their homes and do not have to leave their residences. Inhabitants can renovate their house, but they cannot construct another one. This politics, which limits the increase of houses, poses a great problem to young couples, because they can't have their own house in the park. To resolve this problem, the government constructed outside the park the Run Ta-EK eco-village, where young families can have a new house. The eco-village can receive visitors who can stay at a guesthouse within the village.

The problem stems primarily from the interpretation of living culture. To be alive is to be changeable. Living culture is not fixed like the objects exhibited in the museum. In particular, the growth of the marriage rate is the result of the wishes of the inhabitants. It is a social fact par excellence in the Durkheimian sense of the term. The distancing of the new couple and their parents caused by the law for the protection of the World Heritage opposes the respect for "tradition", family intimacy. Integrity and authenticity, required by The Working Guidelines for the Implementation of the Convention on World Heritage, may enter contradiction with intangible culture. If the rules of the World Heritage institutions must be strictly adhered to, they are becoming more and more of a constraint and the inhabitants cannot escape them. In the Age of Preservation, not only experts of cultural heritage but also inhabitants themselves participate in the constitution of cultural heritage site. But in the case of Angkor, from the beginning, inhabitants were not much interested in Angkor as a cultural heritage; Khmer rouge did not destroy Angkor because they did not give it much value after all. The inscription of Angkor on the World Heritage List aimed at creating a new *lieu de mémoire*, which should constitute a "national" memory. However, Angkor does not seem to constitute a new symbolic center today. UNESCO does not contribute to the creation of symbolic relationship between inhabitants and Angkor, which remains just a source of economic gain. If inhabitants have to be sensitized as UNESCO required, it means the absence of interest in cultural heritage among the inhabitants.

Tourism and age of preservation

Angkor presents another problem. This is about the tourism that develops at the Age of Preservation. Around Angkor, despite the strict constraints imposed on the inhabitants to safeguard and restore the Khmer vestige, capitalist desire penetrates into and around the site. Developers construct hotels and restaurants around the sites and convert World Heritage sites into a space of consumption. A driver of the cab we took complained that the tourism industry was essentially run by foreign capital.

With the intrusion of the tourism industry, the site attracts tourists, and inhabitants begin to work in the tourism and little by little they adapt to consumer society. Developers and inhabitants are still linked to the production and to work. On the contrary, for tourists, world heritage sites are nothing but objects of consumption. Tourists pay the entrance fee and visit the remains. They are delighted with local food at the restaurants located around the World Heritage site. They take pictures and upload them immediately. Soon after, their families and friends see them. Tourists are always connected with their intimate world through the Internet even if they stay in an unknown world. They feel, thus, in security. Tourists and inhabitants intersect and sometimes they exchange. For example, tourists are brought to visit those who lead their lives in the lake. Many come from Vietnam. Their daily life on the water thus becomes an object of consumption.

The example of Lijiang in China shows more explicitly the discrepancy between World Heritage institutions and the Age of Preservation, and it is worth taking a look at.[6] According to the explanation of UNESCO,

> the Old Town of Lijiang, which is perfectly adapted to the uneven topography of this key commercial and strategic site, has retained a historic townscape of high quality and authenticity. Its architecture is noteworthy for the blending of elements from several cultures that have come together over many centuries. Lijiang also possesses an ancient water-supply system of great complexity and ingenuity that still functions effectively today.
>
> (UNESCO, 2020q)

However, after the inscription for the World Heritage in 1997, the old town has much changed. There are bars and restaurants in the city, which attract many young Chinese tourists. This is an unexpected scene for a historical site like the old city of Lijiang. In the bars, singers and dancers play all the night (Figure 5.8).

Naxi people have the tradition of love-suicide. Two young lovers, when they are not allowed to marry choose the death to accomplish the eternal love. This love-suicide was very frequent in the seventeenth century. But, this ancient tradition is transformed into the simple love story, and Lijiang becomes the symbol of love. Lijiang is a famous place that provides much opportunities of encounter. Nightlife in Lijiang is much animated, but the town lost precious silence that it had possessed before the registration for World Heritage List. Moreover, traditional water-supply system is out of use, because accommodations, restaurants, and other touristic facilities wasted much water and then sources were exhausted since some years.[7] The city has, therefore, recourse to sources of the outside to procure water. The water-supply system is one of characteristics of the city. But it was destroyed by the development of tourism after the inscription for the World Heritage List. Because of the development of tourism, most of inhabitants left their hometown. They rent their house to shopkeepers coming from the outside. Inhabitants of the old city became, thus, wealthy. On the contrary, other Naxi cannot benefit from the inscription of the ancient city on the World Heritage List and some are forced to work in the tourism sector. The inscription on the world heritage list then creates an inequality among Naxi people who represent more than 90% of the local population.

The case of Lijiang shows a typical situation of schism between the value of local people, developers, and UNESCO. The value of the old town recognized as World Heritage is not shared by inhabitants, either developers who make the old town exclusively a touristic attraction. There would be an implicit agreement between inhabitants of the old town and developers about their economic benefit, who both ignore the value of Lijiang as cultural heritage. The situation in Lijiang is, in this sense, far from what should be as a World Heritage site; there is no respect for the Guidelines, which requires integrity and authenticity of cultural heritage as necessary conditions of preservation. Houses of Lijiang acquire more commercial value after the inscription on the World Heritage List. World Heritage institutions, contrary to their objective of the beginning, reinforce capitalist desire. The case like Lijiang often happens after the registration.

The prototype of social change brought by the tourism is the example of The Mont-Saint-Michel Abbey in Normandy, France. The abbey has been a high place of pilgrimage since the Middle Ages and at the same time served as the king's prison. After the revolution, it became the prison for refractory priests in 1793 and in 1811, "an Imperial decree transformed the abbey into a reformatory, mainly for common law prisoners and some political prisoners (Abbaye du Mont-Saint-Michel, 2020)". The prison was definitely closed in 1863. The merchants then begged the administration to overturn this decision because there were about 10,000 visitors a year to see prisoners; they were an important clientele. The abbey was classified as a historical monument in 1874, and the fear of merchants disappeared. And since then, the number of tourists has increased (Decaëns, 2013:111). Mont-Saint-Michel and its bay are inscribed on the World Heritage List in 1979. The trace of the prison disappears completely today. It has become a worldwide tourist site (Figure 5.9). There are hotels, restaurants, and

Figure 5.8 Tourists in nightlife of Lijiang, China
Source: Photo by the author.

Figure 5.9 Tourists go to the Mont-Saint-Michel
Source: Photo by the author.

souvenir stores around the abbey. This example shows that a site changes anyway, and it is almost impossible to keep a site in the same state.

Art and cultural heritage

In the case of Lijiang, the way in which houses are used has changed. They have been transformed into a place for tourists like restaurants. In other cases, instead of demolishing the old buildings, they are transformed into art districts. This attempt is different from the traditional norms of preservation. The Guidelines aims to preserve the buildings in the same condition from the time they are registered as cultural heritage. The integrity of the cultural heritage is a prerequisite for registration on the World Heritage List. A cultural heritage should not be tampered with unnecessarily. In practice, however, buildings deteriorate over time and need to be restored regularly but not necessarily with original materials. If so, one of the options is to try to proactively add new elements to the building instead of trying to maintain its original condition. It no longer respects the traditional norms of cultural heritage protection. It provides a new way of preserving cultural heritage that adapts to the Age of Preservation.

This is the case for 798 Art District in Beijing. In 2002, the young artists started their activities by reusing old electronics factories, 798. Gradually, the galleries attracted visitors and today, 798 Art District (Figure 5.10) is becoming a tourist attraction rather than a place of artistic production. Reconstruction using old factories is also taking place in Jingdezhen in China, the city that has a long tradition of ceramic industry. In the 1980s, nationalized factories produced ceramic products. But the government closed the factories and left production to private producers. From then on, various types of production restarted and today, young ceramicists have opened their stores and artistic production such as Bizen ware, presented in the previous chapter, is also developing. Part of the old factories are reused as hotels, galleries, cafes, and meeting rooms (Figure 5.11). And this area is becoming a tourist attraction.

The same phenomenon appears not only in the big city, but also in the countryside. For example, in Naoshima island in Japan, the Art House Project is carried on. The Art House Project gives a new meaning to the traditional houses in Honmura, an area of the island by bringing a different element of contemporary art into the traditional black-walled houses of the area. In this way, the project presents a different value than World Heritage institutions, in which the traditional buildings are not simply preserved as cultural heritage but are inherited by the new concept of the traditional buildings themselves. The Art House Project differs from the so-called cultural heritage in that it combines traditional buildings with contemporary art and actively involves the residents in the creation of the work. The first work of The Art House Project is Sea of Time '98, installed by Tatsuo Miyajima in a renovated 200-year-old house. In this work, a dimly lit room with a water surface, modeled after the Seto Inland Sea, on the floor is decorated with 125 blinking counters. The speed of the blinking counters was freely set by 125 islanders, aged between 4 and 90, who lived on the island. A unique feature of this work is the direct involvement of local residents in the creation of

Figure 5.10 798 Art District in Beijing
Source: Photo by the author.

Figure 5.11 Former ceramic factory in Jinderzhen, China
Source: Photo by the author.

Figure 5.12 "Dentist" by Shinro Ohtake in Naoshima
Source: Photo by the author.

the artwork, rather than simply displaying a new work of art in an old house, and in 2018, 20 years after the work was created, a time setting meeting is being held to re-set the speed of the counters. Another work of art is an old dentist's house (Figure 5.12). The artist put on the wall of the house an old signboard to recall the past.

Society inevitably changes. It means the actions of members of society change and their way of seeing the world as well. The space they produce is also transformed. The rigid norms of cultural heritage institutions no longer fit the Age of Preservation. The examples of the 798 Art District and Naoshima show an attempt to adapt buildings that are no longer in use to social change by giving them artistic value. It is a new way to preserve cultural heritage. The way of preserving cultural heritage changes with social change.

Catastrophe and cultural heritage

The phenomenon that brings radical change par excellence is catastrophe. When an already-valued cultural heritage site is destroyed, it is regarded as a significant loss and urgently sought to be restored. For example, after the 2015 earthquake in Nepal, BBC web news reported that as well as the devastating loss of human life in Saturday's earthquake, Nepalis, and the world, have lost parts of the country's unique cultural heritage. The article titled "Nepal's Kathmandu valley treasures: Before and after" appeared just after the Nepal earthquake 2015 and showed pictures of World Heritage sites before and after the earthquake to demonstrate the destruction of monuments of worldwide importance in Kathmandu (BBC, 2020).

UNESCO was more anxious, and the Director-General of UNESCO immediately expressed her concern about the destruction of the Kathmandu temples, a world heritage site, by the disaster, and this attitude is the traditional administration for the protection of cultural heritage.

> UNESCO has mobilized its expertise as well as international support for Nepal at this difficult time, including for the safeguarding of its heritage. UNESCO is currently preparing to field an international expert mission to undertake an in-depth damage assessment and, based thereon to advise and provide support to the Nepalese authorities and local communities on its protection and conservation with a view to recovery.
>
> (UNESCO, 2020r)

Cultural heritage is as much important as human life. At least, the collapse of world heritage monuments is a great loss for "the world". BBC and UNESCO innocently believe that they have the right to worry about the destruction of monuments in Kathmandu valley. However, for the victims of the disaster, that is not the most urgent concern. They are anxious about their future life. Their interests lie in the reconstruction of their residence and not in the restoration of World Heritage sites. Even if they think of the restoration of monuments, this is not because they recognize temples of Kathmandu valley as cultural value but because they believe in the religious meaning.

In fact, according to Mircea Eliade, at first people fixe a "sacred center" such as mountains inhabited by deities, temples, and palaces. Places such as mountains, temples, and palaces are thought to become sacred centers because they stand at the intersection of heaven, earth, and the unknown. The center is seen as a place linked to the world of the unknown, a world inhabited by the ancestors and spirits. At such places, the living can symbolically interact with deities and ancestral spirits. Kathmandu valley is one of these places controlled by the symbolism of the center (Eliade, 1969). From this point of view, temples are not regarded as having an aesthetic or historic value. The reconstruction of temples damaged by the disaster should aim to crystalize the memories of those who died unexpectedly. In this sense, the reconstruction is an act of mourning for the dead. It helps, thus, mitigate the grief of those who lost their loved ones and to symbolically overcome the difficulty insofar as those parties exist in the hereafter. The reconstruction of damaged temples allows the recreation of a center, an intersection of the living and the dead. This is precisely the order of recollection.

On the contrary, the philosophy of World Heritage regards temples as exclusively cultural monuments. According to the description by UNESCO,

> these monuments were defined by the outstanding cultural traditions of the Newars, manifested in their unique urban settlements, buildings and

structures with intricate ornamentation displaying outstanding craftsmanship in brick, stone, timber and bronze that are some of the most highly developed in the world.

(UNESCO, 2020s)

Descriptions like "unique urban settlements", "outstanding craftsmanship" present only their artistic and architectural values. As such they have to be preserved for evermore. Certainly, World Heritage institutions legitimize a new tendency which preserves as shared heritage the various experiences of people and the culture that they have created. However, these experiences and culture are not religious. They regard them only as cultural heritage.

If UNESCO's point of view toward the reconstruction of the World Heritage site in Kathmandu is traditional, showing the effect of the Age of Preservation is the case of Great Hanshin-Awaji Earthquake. It is the attempt to make the Nojima fault a natural heritage. Immediately after the great Hanshin-Awaji earthquake, active fault researchers began investigating the fault in Hokudan-cho next to the residents of the severely damaged Hokudan-cho, who were living in evacuation centers. Thanks to these considerable efforts of the researchers, it is only three years after the earthquake that it is classified as a natural heritage site. This is an unusually short period of time. The Hokudan-cho Earthquake Memorial Park opened in the same year. From then, the fault is exhibited at the Nojima Fault Preservation Museum (Figure 5.13). A house that was crossed by the Nojima Fault is preserved as a memorial house with the fault.

At first glance, the concept of the Nojima Fault Preservation Museum is also based on the same principle as World Heritage institutions. The website of the Hokudan-cho Earthquake Memorial Park states that "lessons learned from the past will be passed on to the future;" "At the Nojima Fault Preservation Museum, we preserve the Nojima Fault as it is, so that visitors can feel the horror and threat of earthquakes and learn the importance of earthquake preparedness;" and "we transmit the lessons learned from the earthquake and the awareness of disaster prevention. The lessons of the past need to be connected to the future" (Nojima dansô hozon Hokudan kinen kôen, 2020). This is an attempt to pass on the events of the "past" to the "future", just as the Louvre is preserving and exhibiting the Mona Lisa for future generations. The designation of the Nojima fault as a natural monument is similar to this concept. In order to preserve the state in which the movement of the active fault is clearly visible, conservation measures are necessary. Otherwise, scientific evidence of the movement of the active fault, which is a natural phenomenon, will be lost forever. Moreover, even if the Hanshin-Awaji Earthquake was a one-time event, it does not mean that the disaster will not happen again in the affected areas. Therefore, the museum will function as an educational device that teaches the "lesson" that we should not always be negligent in preparing for natural disasters from which we cannot escape, by arousing the fear of the disaster. The Nojima Fault Preservation Museum is based on moral values in the broadest sense of the word: "lessons learned" and "disaster preparedness".

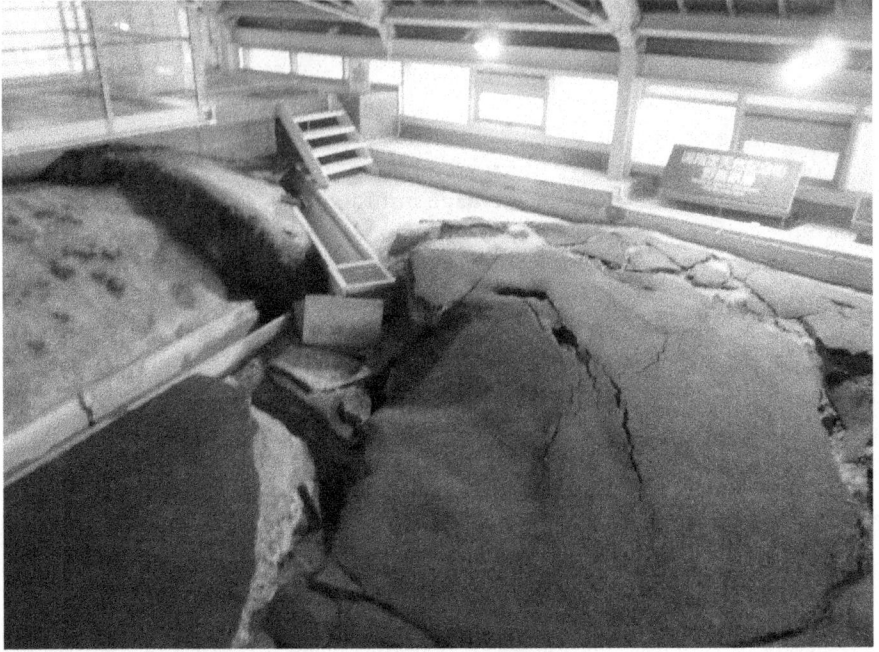

Figure 5.13 Active Fault preserved in Awaji Island
Source: Photo by the author.

However, the act of preserving the damaged house as a memorial house and exhibiting a reconstruction of the disaster is different from the old way of protecting cultural properties. Aesthetic value is absent from the Nojima Fault Preservation Museum, and the museum's collection, which presents the facts of the disaster, does not have any inherent value as a work of art. On the contrary, for the victims, they are nothing more than exhibits that are linked to their abhorrent memories. This is a negative legacy. The Nojima Fault Memorial House is not a historically valuable building like the Kathmandu temples. The memorial house gained its value because it was damaged and because an active fault crossed it. It is only as a result of the occurrence of the new disaster that it becomes an object of preservation. It is not a cultural heritage that has withstood the winds and storms of history, but one of which preservation is determined immediately after the earthquake based solely on the criteria of the "present". Historical evaluation is not left to the future, but the value of the present is absolute.

This is not only true in Japan, but also in the Aceh Tsunami Museum established five years after the 2004 tsunami and the National 9/11 Memorial Museum established ten years after the 2001 terrorist attacks in the United States. In Banda Aceh, the boat that reached onto the roof of a house at the time of the tsunami is preserved (Figure 5.14). These examples can be seen as absorption of the past into the present. Rather than locating things in the past in a linear time

Figure 5.14 Kapal di atas rumah, The Boat on the roof after the great tsunami in Banda Aceh, Indonesia

Source: Photo by the author.

sequence, they grasp the events of the past as something that could happen in the present. By preserving buildings and other objects that show the tragic events of the past, they arouse fear and encourage people to be aware of the fact that their present safety could be threatened at any moment. At the same time, however, it also tries to show that safety can be maintained by having an awareness of disaster prevention and crime prevention.

Consumer society and Age of Preservation

Another characteristic of the Age of Preservation is that it goes hand in hand with the arrival of the consumer society.

Sociological literature observes transformation from the logic of production to the logic of consumption. In the early days of industrial capitalism, the logic of production presupposes effort and patience. Discipline is required in the work-place. But capitalism has another tendency, which aims at accelerating production to make more profit and creates the desire to consume as soon as possible. Capitalist system, thus, seeks to absorb the future into the present. The future is not a later time. It is not to come, but it must be built even now. An example is the loan system. It makes it possible to make big purchases even if there is a lack of cash. Consumers buy an apartment or a car that they would not have gotten without the loan system. It progressively reduces the distance between the present and the future and at the end there is no more lap of time between the present and the future. Global capitalism accelerates reduction of temporal and spatial distance, and there is more and more mobility. David Harvey calls this irreversible global change "time–space compression". Consumers are constantly looking for novelties in the consumer society in which time is compressed. Fashion is a good illustration of the time–space compression. It is becoming normal in this type of society that what was fashionable yesterday is no longer fashionable today. Harvey considers *volatility* and *ephemerality* as characteristics of postmodernity.[8] He says, "the first major consequence has been to accentuate volatility and ephemerality of fashions, products, production techniques, labor processes, ideas and ideologies, values and established practices" (Harvey, 1989:285). Ephemerality is an observation of a state in which everything changes from one moment to the next and the direction can hardly be predicted. This remark of Harvey is close to the reflection of Zygmunt Bauman. According to Bauman, the reduction of the duration is a characteristic of the age of liquid and light modernity. To be transient becomes a value and duration is less and less important (Bauman, 2000:125). Bauman and Harvey agree that ephemerality appears with the expansion of global capitalism and the consumer society.

In this circumstance, the individual is liberated from familial framework. Jean Baudrillard gives an example by analyzing the use of furniture in family and society.[9] "The typical bourgeois interior is patriarchal". "The furniture is arranged around an axis that ensures the regular chronology of conduct: the always symbolized presence of the family itself" (Baudrillard, 1968:19). But this family moral order is replaced by the functional order that leaves more freedom for the arrangement of the living space. Property becomes more personal and customized. As a consumer, the individual conceives the organization of space and makes use of his/her property freely. Individual-consumer accumulates objects-merchandises but these objects do not become familial or common legacy. And if objects are used, they are usually thrown out, except objects from the personal collection. Individual-consumer lives ephemeral world essentially mediated by objects and loses the notion of continuity that family moral order provided. Furthermore, in a consumer society, commodity ceases to satisfy needs and instead become signs. A consumer society automatically generates commodities-as-signs in its system of signs, which makes no reference to reality.[10] (Baudrillard, 1970).

If the merchandise is a sign, the producer gives it a meaning and through it, communicates with the consumer. The consumer, for his part, gives a meaning as he pleases. For example, the way of arranging the interior of the house to one's liking. Consumers live fashion as a theater where they can act as actors. They are, thus, used to seeing the world through the presence of commodities. Consumption is not an act of buying and owning a commodity but is an integral part of existence itself. Humans in the consumer society consume not simply for their material needs but to prove their existence. This way of grasping the world, generalized in the consumer society, also applies to the way of seeing monuments or the landscape. They are also commodities. A World Heritage site is a sort of commodity and tourists consume it. People search on Twitter or Instagram for the place to visit, and they show pictures they have taken on social media as well. This whole series of phenomena is called the commercialization of culture (Greenwood, 1989:173). The cultural act becomes the act of consuming.

Consumer society and Age of Preservation have the same characteristics. Ephemerality or transience, that is, the reduction of duration, means the absence of history and of distinction between the past and the present. As there is less distinction between experts and non-experts in the selection of cultural heritage, the inequality that once existed between producer and consumer as well as seller and buyer is being reduced. The consumer can acquire more information and give a quotation to a good he/she has bought on the Internet. These similarities do not happen by chance. The Age of Preservation is both the result of the consumer society and the promoter of its development.

Iconoclastic ideology

Even if the preservation of cultural heritage is imperative at the Age of Preservation, there are still elements that oppose it. The first factor is the existence of ideologies that do not accept World Heritage institutions. Recently, two groups of Muslim fundamentalism based on iconoclastic ideology have attacked World Heritage sites. The first group is the Taliban, which destroyed the statues of Buddha in Bamiyan valley. The second one is Islamic State, which attacked the Site of Palmyra in Syria. Both of these targets of attack are classified as World Heritage. The attitude of Taliban and Islamic State is the total refusal of the logic of World Heritage institutions in which most valuable monuments in human history should be preserved. Paul Veyne, French historian of antiquity, complained about the destruction of the Site of Palmyra by Islamic State and stated that despite his advanced age, it was his duty as a former professor and a human being to convey his amazement in front of this incomprehensible devastation (Veyne, 2015:10). For those who are sensitive to the history of humanity and its heritage, the destruction of the Roman vestige is a serious attack on the history of humanity. Precisely, Veyne emphasizes his responsibility not only as a former professor but also as a human being. As such, the historian feels obliged to write a book on the magnificence of Palmyra lost forever. However, it is not known to what extent Veyne's concern was shared. Indeed, the notion of history of humanity seems to be fading little by little. In consumer society, a cultural heritage site is just a tourist attraction. Its images circulate on the Internet as potential objects of consumption. Therefore, there is no place for the history. World Heritage institutions entirely based on the history, which presupposes the linear conception of time and the notion of progress, are losing their sense at the same time. On the other hand, we do not know how much interest the inhabitants have in safeguarding the World Heritage site in a state of war, especially since it is difficult to link the vestige of Roman antiquity to the daily life of Muslims.

Culture of oblivion

Another element that opposes the Age of Preservation is culture of oblivion. In Europe, preserving cultural heritage means preserving material objects and tangible cultural properties. Even objects that have deteriorated, such as the ruins of ancient Rome, for example, can be preserved as elements of cultural heritage as long as the remnants are preserved. There is beauty to be found in such ruins. Nevertheless, there are societies that do not only need to preserve but also value forgetting. In Japan, the ruin is rather an object of fear. In Kenji Mizoguchi's film *Ugetsu Monogatari* (*Tales of the Pale and Silvery Moon After the Rain*), a ghost appears among some ruins and kidnaps a living person. This ghost is the spirit of a person who was not given a proper funeral. Likewise, legend has it that when an item that has been used for a long time is carelessly discarded, it turns into a ghost. Some people therefore feel somewhat fearful of museums, with their rows upon rows of out-of-use objects. This fear is derived from the sense that these obsolete objects have been stockpiled rather than given a proper funeral. Moreover, in Japan, traditionally, there exists a belief that used objects have a bad spirit which should be exorcized in one way or the other. *Hari kuyō*, a sort of requiem service for broken needles, is a typical example of the exorcism. The order of recollection makes it possible to forget the calamity by praying to deities. Forgetting is thus an integral part of the order of recollection.

The culture of oblivion is, in fact, what is radically opposed to the Age of Preservation because it voluntarily seeks to forget. Diversity pleaded in the Nara Document cannot give a place for the act of forgetting or culture of oblivion. Culture that denies the preservation of heritage is completely excluded. Therefore, a fundamental paradox appears. As long as heritage institutions deal with very limited historical objects and monuments, the culture of oblivion is left aside and therefore intact. But as soon as the living culture itself is an object of control through cultural heritage institutions, it risks disappearing, whereas many societies possess the culture of oblivion. In fact, intangible culture that UNESCO suggests respecting today contains elements of culture of oblivion. But when UNESCO gives respect to traditional culture, the importance of forgetting is completely ignored. This is because the Age of Preservation does not accept forgetting.

In the Age of Preservation, what was traditionally not part of the field of cultural heritage can be considered as such. The house above the Nojima fault that caused the earthquake is preserved. It is the event that gave value to the house like the others before the earthquake. On the other hand, the cultural heritage becomes an object of consumption, and visitors consume information about the event like a movie scene. Capitalist desire invades the sphere of museological desire with the development of consumer society, and frontiers between museological desire and capitalist desire become blurred. Museological desire and capitalist desire, which were strictly separated before, are interpenetrating, and the latter absorbs the former desire.

We can state that the so-called "culture" somehow absorbs both the museological and the capitalist desire. The "cultural desire" that makes everything cultural appears whereas museological desire defines the distinction of cultural heritage and is therefore very selective. Cultural desire blends well with the desire to consume. An object can become both a commodity and a cultural object. This state of culture everywhere can be called "cultural ubiquity". If everything that exists has a cultural meaning, then everything has the right to be remembered and never forgotten. But from another point of view, cultural ubiquity runs the risk of making cultural value meaningless, since if everything is cultural, none of it makes sense.

Notes

1 Nevertheless, Lévi-Strauss himself seems to see in the objects exhibited in museums a purely symbolic or almost mythical function like objects in cold societies.
2 "Some misinformation was provided by a number of sources including a few politicians, less informed journalists and the general public, often not aware of complex nomination and monitoring processes under this international legal instrument" (Ringbeck and Rössler, 2011:205).
3 A Japanese researcher interviewed an official who was dealing with this case and, according to him, there was an error in this description. It is not 5 km but 3 km from the center of Dresden (Nanazawa, 2010:37).
4 When UNESCO describes the inhabitants as holders of "archaic cultural practice", its description is not far from evolutionism of nineteenth century. The autogenous development required by UNESCO is not really autogenous.
5 According to a responsible of APSARA, in spite of this politics of control, those who illegally settled in the park don't stop increasing.
6 The Old Town of Lijiang is located on the Lijiang plain at an elevation of 2,400 meters in southwest Yunnan, China (UNESCO, 2020q).
7 We did our research in 2017.
8 Nevertheless, in the case of Angkor, for example, ephemerality lived by inhabitants is not same as that of tourists. Cambodian people had experiences of suffering from Khmer Rouge regime. At that time, they did not know what would happen the following day. In this sense, they were living an ephemeral life but in a completely different way. Their conception of ephemerality is more related to the incertitude while foreign tourists live that of consumer society in security.
9 Baudrillard focused his sociological interests on the relationship between human being and objects, while in classical sociology, human relationship was the main object of research. He inaugurated, thus, sociology of objects.
10 Baudrillard (1970) contended that the process of consumption can be analyzed from two standpoints: the process of signification and of communication and the process of classification and of differentiation.

6 Frozen time–space and cultural heritage

If the Age of Preservation accelerates paradoxically the annihilation of cultural value, can cultural heritage institutions contribute in any way to something? And if the contribution is possible, what will it be[1]? These questions must be answered in relation to time and space. They are, first of all, related to the time because in the Age of Preservation the social perception of time is radically transformed. The boundary between past and present is removed, and there is less distinction between the two whereas originally, cultural heritage institutions presuppose a duration. And then, the question of the space must be analyzed insofar as World Heritage institutions make extensive use of spatial concepts such as territory and landscape. Furthermore, the political and social management of space becomes more important even more so as the absence of temporal duration causes the relative lowering of the temporal dimension's weight in the social control. Finally, the theoretical constitution of the time–space relation must be envisaged. As we shall see, the time–space relation is the most important issue to understand the possible social contribution of cultural heritage.

Time and vector

To understand the change of collective consciousness of time, we introduce the notion of vector. We suppose three types of vector, respectively future-oriented vector, past-oriented vector and zero vector. Chapter 1 has shown that, initially, the cultural heritage institutions were based on the notion of progress in history as they consist of transmitting valuable objects and monuments to future generations. The Louvre was established for teaching and learning evolution in arts, which symbolizes the progress of humanity. The vector is here oriented toward the future. In contrast, Chapters 2–4 have implicitly dealt with the existence of the past-oriented vector. In Chapter 2, the trauma caused by the atomic bombing was evoked. Trauma is the expression of negativity toward the past. Inscription on the World Heritage List cannot easily remedy trauma or negative memory. On the contrary, it even happens that the conflict situation arises, as shown, in Chapter 3, by the inscription of Hashima Coal Mine on the World Heritage List. Another example of a past-oriented vector is the case of nostalgia. It is the positive feeling toward the past. In the case of nostalgia, inscription on the World

DOI: 10.4324/9781003014904-7

Heritage List aims to conserve monuments that arouse nostalgia by considering that without inscription, the monuments would one day disappear and so would nostalgia. Chapter 4 highlighted the logic of actualization, the logic underlying the idea of intangible cultural heritage. This is also the vector oriented toward the past. The actor and his/her spectators can thus themselves return to the past.

Trauma, nostalgia, and logic of actualization are all related to the past-oriented vector. However, the vector that is once oriented toward the past can change its orientation to the opposite direction. The typical example is the logic of actualization. Indeed, actualization is the attempt to resurrect in the present that which once existed or is believed to have existed in the past by performers or artists. The vector of this logic goes from the past to the future when the past is resurrected by artists in the present. For example, an actor somehow brings the past back into the present when he/she plays on stage. The same logic also applies to tangible objects and monuments. In the case of the Historic Centre of Warsaw, for example, the World Heritage site is the reconstruction of the city before the Nazi destruction. It is considered actualizing what would have been before the war. The positive memory of the city has been resurrected. The vector is once oriented toward the past, which is the city before the destruction. Then, when this past is restored in the present, the vector goes toward the future. This change of the orientation can be called "vector rotation". The vector rotation takes place at the time of the inscription on the World Heritage List; World Heritage institutions ensure the transmission of the site toward the future. In the case of Hiroshima, the past experiences of suffering not only remain but the survivors still suffer and the suffering will continue. The cursed past keeps coming back. The trauma suffered by the survivors, their families, and Hiroshima citizens always orients their conscience toward the past although they want to escape it. The movement for the inscription of Genbaku Dome on the World Heritage List was an attempt to turn this past-oriented vector to the future to overcome the suffering. The World Heritage institutions can give rise to the vector rotation.

In all societies, the future-oriented and past-oriented vectors are present and functioning more or less. When the opposite vectors have the same volume, the state of the zero vector appears. Consumer society allows to realize what would have been possible only in the future and therefore to absorb the future in the present by means of the loan system. The vector here is oriented from the future to the present. When the loan system is well established, the question of the orientation of the vector no longer arises. There is only the here and now. This is the situation that can define the zero vector. Immediate consumption and the system that enables it constitute the zero-vector situation. The notion of ephemerality or transience suggested by Harvey and Bauman is precisely the expression of this situation. The zero-vector situation seems intact if nothing happens. But it is difficult to ensure this condition always intact because of the permanent underlying risk. When, for example, a great earthquake has occurred, it reveals the existence of uncertainty. The "double loan" is a typical example of uncertainty created by a catastrophe. Many victims lost a house or apartment that had been purchased on loan. Even after the loss of their property, they are still obliged to repay the loan. If they want to buy another house or apartment, they must obtain another loan, thus resulting in a double loan. This is what happened, in Japan, after the great Hanshin-Awaji earthquake in 1995. Of course, not everyone can make a loan purchase. The bank will examine the income and social status of the person who wants to make the loan. This person must prove his/her ability to manage his/her life. The image of the future is clear and the trajectory of his/her life must be well defined. If the future is rationally planned in advance, it is no longer the future itself. It is already in the present. A great disaster breaks this rational way of life. A big earthquake also brings survivors to conceive the changing character of the world. The present is considered as a moment inevitably disappearing.

The upper class tries to take action against risks of different kinds. For example, to avoid all risk, Heritage Park constructed in South Africa "is to be entrusted to hidden TV cameras and dozens of hired gun-carrying guards checking passes at the security gates and discreetly patrolling the streets" (Bauman, 2000:93). Nevertheless, even the upper class cannot fully prepare for all the risks, as the example of COVID 19 shows. The feeling of uncertainty permeates the population. This is why the consumer society is also considered the risk society or the surveillance society. Since there are always potential risks, there is always a need for surveillance. Risk and surveillance are therefore two sides of the same coin.

Modernity and postmodernity

Three types of time vector can be explained more deeply by referring to the philosophical and sociological literature on time.

First of all, the notion of progress based on the Enlightenment that has played a decisive role in the constitution of cultural heritage institutions belongs to the category of the future-oriented vector. In the nineteenth century, the notion of progress was attached to evolutionary time. It gave rise to historical consciousness and gave value to history. At the same time, it dissipated the tendency to hold on to the past, and the past-oriented vector seems to be annihilated. The valorization, even the necessity of history is, according to Pierre Manin, the characteristic of modernity. To be modern is to be historical (Manin, 1994:13). A modern person is fully aware of his/her historical existence, of living in history. Modern philosophy sees history as the process of the liberation of mankind. Modern awareness positions, thus, the past that has been surpassed by the present. In the modernist view, the past is defined by the present and the present is defined by the future. Time is irreversible.

Jean-François Lyotard is one of the philosophers to question the concept of modernity taken for granted. According to Lyotard (Lyotard, 1988), modernity defines the series of events grouped in the name of the universal history of humankind. He holds that modernity must be formed under the philosophy of liberation, that is, the perception that humankind is being liberated. Of course, this philosophy of liberation – "the Grand Narrative" – can be framed in many ways, liberation from exploitation and alienation under Marxism, liberation from poverty by technology and industry under capitalism; and redemption through the love of Christ are all modern reiterations of the philosophy of liberation. Modern awareness positions the past as a "passé" age that has been surpassed by the present. However, according to Lyotard, the predominance of Western European thinking in modern philosophy, which has been dominant for two centuries, is beginning to crumble. This was triggered by Auschwitz. If history was heading toward the liberation of humankind, then modern philosophy has no answer for the question of how was it that so many Jews were allowed to be massacred. Thus, while the perception that modern philosophy is powerless extends, the hidden side of modernity is being gradually exposed. To understand this side, Lyotard introduces the concept of postmodernity.

How, then, should we regard postmodernity? Using the example of a modernist painting, Lyotard said that a painting can only be modern by first being postmodern. For instance, Cezanne sought to create a new art movement by rejecting impressionism. That could be said to be a postmodern move because it came after impressionism, a modern movement. However, for artists like Braque and Picasso, who came after Cezanne, Cezanne was already modern. So, whether it be Cezanne or Picasso or even Duchamp, who came later still, they first had to arrive on the scene as postmodern artists and remain faithful to the modernist philosophy of "question everything that came before" in order to be perceived as modern artists. If this is so, then their elevation as modernist masters was consequential rather than intentional. Given that, according to Lyotard (Lyotard, 1988:28), "postmodernism is not the end of modernism but the process by which modernism is born".

Thus, postmodernism, as has been often pointed out, does not refer to a new zeitgeist that follows on from modernism; indeed, the very idea that a new zeitgeist must follow on from modernism is a modernist view of history. Seeing postmodernism as simply a new zeitgeist is to deprive the concept of its defining characteristic. Indeed, whereas modernity gives meaning to the future by defining history as the process of human liberation, postmodernism dismantles the historical view and seeks meaning or to express the very process by which meaning is generated. In the Lyotardian philosophy, in a somewhat schematic way, there are two perceptions of time. One is that of modernism, in which history is seen as the process of the liberation of mankind. The second is the postmodern view, which moves away from that kind of past–present–future, straight-line perception of time. Even if modernist view and postmodernist view seem opposite, for modernity to exist, it requires not just the modernist historical view, which draws the schematics of human liberation, but also the postmodern perspective, which questions those plans. The postmodern perception does not necessarily oppose the modern view of history. Modernity, after all, can be seen as having both modern and postmodern aspects and would not exist without both of them.

If postmodernity is apprehended as such, postmodern works do not seek to be evaluated in accordance with pre-prepared standards; they are a quest for new standards of value. In that sense, the very act of unveiling a new piece of work takes on the appearance of an event in itself; postmodernism is the tendency toward the chaotic and lies beyond the bounds of established standards. Lyotard compares the postmodern attitude to the psychoanalytical process. In psychoanalytic therapy, the patient seeks to find the hidden meaning of his/her disorder by freely associating apparently inconsistent elements with past situations (Lyotard, 1988:119). Similarly, postmodern artists try to discover another sense repressed by modernity (Ogino, 2015).

In this attempt, the linear history of art no longer makes sense, and the past and the present are confused. Postmodern architect intentionally confuses styles that have existed in different periods without giving a sense of progress. Robert Venturi celebrated both "the ambiguities, inconsistencies and idiosyncrasies of the Mannerist and Baroque architecture of Rome", and "popular culture and the ordinary architecture of the American Main Street"[2].

The philosophical concept of postmodernity introduced by Lyotard reveals the existence of the vector hidden by modernity, which only sees the vector oriented toward the future. The inverse vector, a vector oriented toward the past, is explicitly recognized. It reflects a transformation that took place in the 1980s. In fact, the interest in memory increased from the 1980s. The concept of collective memory was taken up by Pierre Nora (Nora, 1984). Even negative memory was brought to light. In addition to the concept of trauma discovered by Sigmund Freud, the concept of PTSD was established by the American Psychiatric Association in 1980. If, as Manin pointed out, linear and future-oriented history is the pivot of modernity, all these phenomena clarify the existence of the vector that goes in the opposite direction of the one intended by Modernity. Modernity is thus no longer uncontested. It has begun to be questioned.

Sociology and modernity

Sociologists seek, like Lyotard, to understand the global transformation that occurred in the 1980s. But they do not have the same perspective as Lyotard. For example, Giddens, who holds modernity dear, points out that the

> emptying out of progress by continuous change, are so different from the core perspectives of the Enlightenment as to warrant the view that far-reaching transitions have occurred. Yet referring to these as post-modernity is a mistake which hampers an accurate understanding of their nature and implication.

And he stated "we have not moved beyond modernity but are living precisely through a phase of its radicalization" (Giddens, 1990:51).

Sociological literature insists on a feature of modernity; historical consciousness strictly distinguishes between before and after. This distinction is essential in sociology that emerged from the nineteenth century onwards. Ferdinand Tönnies proposed the distinction between *gemeinschaft* and *gesellschaft*. Somewhat similarly, Durkheim distinguished between mechanical and organic solidarity.

This distinction introduced by the classical authors of sociology led to the concepts of traditional and modern society. Even today, the concept of modernity becomes central in sociology. For example, Anthony Giddens insists on the discontinuity of modernity.

> The modes of life brought into being by modernity have swept us away from all traditional types of social order, in quite unprecedented fashion. In both their ex-tensionality and their intensionality the transformations involved in modernity are more profound than most sorts of change characteristic of prior periods.
>
> (Giddens, 1990:4)

Modernity is an equivocal concept that consists in covering the whole of the way of thinking and acting. But whatever the content given to modernity, the only common point is the insistence on the great transformation it has created. The transition from traditional to modern society is, thus, privileged in sociological theory, and the distinction between the two types of society becomes a postulate of sociology. This paradigm of sociology mainly takes into account the future-oriented vector.

In order to insist on the radicalization of modernity, the term "late modernity"[3] is used by sociologists such as Ulrich Beck, Hartmut Rosa, and Giddens himself. Giddens calls the nature of late modernity reflexivity. Beck does the same, but he insists in particular on risk. Rosa, on the other hand, sees acceleration as characteristic of late modernity. But in fact, these sociologists of late modernity are aware of the vector oriented toward the past, and they translate it by the term reflexivity and reflexive modernization. Reflection implies two processes. In the first step, reflection requires always reflecting on what has happened and a regular return to the past. In the second step, from this reflection on the past, a decision is made on what to do in the future. In this phase, the vector returns to the future again. Reflexivity is thus a conscious act of rotating the vector. And in any case, late modernity clarifies what is hidden by early modernity, the same element revealed by the postmodern.

Sociology of collective memory

Long before the appearance of the concept of reflexivity, Durkheimian sociology had noticed the past-oriented vector. Durkheim, in his later studies of religion, tried to clarify the elementary form of religious life. He inferred from data on small-scale societies a kind of common structure of human religious life. The structure that existed in a small Australian society in the past has not disappeared. On the contrary, it exists in all societies as a common structure. Durkheim somehow resurrected the past in order to understand the present of religious life. His method therefore includes a vector oriented toward the past (Durkheim, 1912).

His disciple Maurice Halbwachs took this method further by being the first to make memory the object of sociological studies. His argument focuses on the way in which an individual constitutes his/her memory in relation with others.

He points out that when an individual remembers, he/she relies on a social framework. Social framework gives opportunities to remember the past related to one's individual life. Some members of the group and/or some important events constitute a framework which offers a reference point for remembering (Halbwachs, 1925). Memory defined as such is not well demarcated. It is something uncertain and difficult to comprehend. Place and object as framework of memory, thus, play an important role in reducing this uncertainty of memories through its materiality. Furthermore, memory changes with time, because those who compose the group change. If the outside of the group influences some members of the group, the group reacts in various ways. The group can accept new elements based on memories of outside openly or in taking the former elements into account. The group can also reject completely new elements to defend its traditions. Conflicts, thus, can occur concerning collective memories, and the reaction of the group can give rise to a change. Collective memory is closely linked to the living past of those who compose the group. If the past loses this vivacity for the members of the group, it will be outside of the memory.

On the other hand, Halbwachs remarks, in particular in his study on topography and religious memory, that knowledge of what was original is secondary, or even unimportant (Halbwachs 1941). Collective memory conceived in this way comes close to collective representation, or even becomes synonymous with collective conscience when the religious group is well established. At this stage, religious memory is no longer mutable. It is considered as fixed once and for all by the group. For example, the Church has its own memory, which is religious doctrine. It seems to the members of the Church that the doctrine is unchanging. Christians believe that the memory of Christians faithfully keeps the life of Christ. From an external point of view, the doctrine has changed. But believers firmly believe that it is immutable. In the religious group that is well established, the vector is definitely oriented toward the past. This past becomes, thus, the primordial memory.

A new value under the guise of an old one

On the contrary, in a circumstance where the established order is in crisis and it is not yet clear whether a new group prevails over the old one, the new group sometimes uses ancient elements that are almost forgotten. New values appear under the guise of old but forgotten values. This is the resurrection of the past (Halbwachs, 1925:183). It "projects into the past the conceptions it has just developed" (Halbwachs, 1925:185). Even if the conceptions are new, they must sometimes look old.

The rise of a nation-state is precisely the case. Cultural heritage institutions constitute a principal means to promote the history of the nation, especially in the case of Western Europe. History chooses important events and tries to give these events an order to simplify them for the didactic purpose. But history is not enough for the constitution of the nation-state. The constitution of collective memory also plays a role. Collective memory here, in the sense of collective representation, is different from history even if history can influence the formation of collective memory, because collective memory holds only a part of the past that is still alive in the conscience of the group. In other words, during the formation of the nation-state, there are two vectors that come into play, the future-oriented vector and the past-oriented vector. There are societies in which the future-oriented vector prevails, like Western Europe. There are others in which the past-oriented vector is rather privileged, like Japan.

In fact, regarding the modernization in Japan, the lack of historical consciousness is pointed out. For example, the political scientist Masao Maruyama (1983) remarked research on the history of Japanese thought could not thrive because Japanese society was founded on a bedrock not strong enough to withstand a historical axis of examination. This axis, according to Maruyama, is created through subjective objectivation of, and full reconciliation with, the past. Maruyama assumed a modernist, straight-line view of history for his axis model. Western thinkers were constantly attempting to challenge and sublate prior thought, so it is simple to follow the historical changes in Western thought. But this is not the case for Japan. Japanese people's spiritual lives lack such an axis and, therefore, as Maruyama (1983:12) says, "a variety of philosophies, which have become entrenched through a certain degree of chronological order, simply affect a spatial rearrangement within the spirit and tend toward timeless coexistence, thus losing their historical structure". The lack of a historical unifier that enables the sequencing of philosophy inevitably creates a situation where philosophy can only exist as a variance without unity. This, therefore, precludes the creation of the new, overcoming philosophies through the conflict of new and existing philosophies because such conflict cannot occur (Ogino, 2013:101–104).

According to Maruyama, "forgotten" philosophy which has been pushed to the side or relegated, and therefore disappeared from our awareness, is only temporarily swept to the further recesses of our minds and sometimes floods back into the forefront of the minds in the form of memories. This flooding back of memory "often presents itself in the form of language, where a person who has long become used to not using their native dialect suddenly reverts to their native patois in moments of unguarded surprise" (Maruyama, 1983:13). Maruyama uses the poet Kôtarô Takamura as an example of this phenomenon. Upon hearing of the outbreak of the Pacific War, Takamura wrote:

Yesterday became the distant past
And the distant past became the now.
"The Emperor is in danger"
This few words
Decided all for me.
Mother and father were there.
The mists of my boyhood home
Filled the room.
The cries of my ancestors filled my ears,
Gasping, "His Majesty! His Majesty!"
And my consciousness was blinded by the brilliance of the moment.[4]

The crisis of the outbreak of the war suddenly brought the boyhood memories that Takamura had forgotten hurtling back and rallied them into imperial ideology. The poet remembers his ancestors through the emperor and the crisis of the imperial system. In other words, the crisis of the nation is also experienced as the crisis of his family. Memories about ancestors are hidden in ordinary times but are recalled at a given moment, in this case, the threat of war. The mechanism, in which Takamura remembers his childhood, his parents and his ancestors is similar to the imperial system. Imperial ideology is based on the conception of time different from the linear conception of time because it consists in resuscitating ideal antiquity to redress the social order in peril. Instead of emphasizing the break between the past and the present, the origin is idealized, and this ideal must be resuscitated. The vector is first of all oriented toward the past and then turns to the present in which the emperor is in danger.

However, this excited feeling in the face of threat does not last long. After the war had ended and the aforementioned crisis had abated, Takamura once again reverted to his "memories" of Rodin. At first glance, this appears to be a turnaround, but it is simply a case of Takamura having so much unsequenced schooling inside of him, so that it is no wonder that different parts of his learning came to light in reaction to different situations. There was only "a rearranging of things that were lurking somewhere to bring them into the light" (Maruyama, 1983:13).

From Maruyama's argument two observations can be made. First, in the formation of the nation-state, the past-oriented vector helped to build the new social system; the vector rotation precisely occurred. The formation of the political-ritual institutions of cultural heritage protection follows the same logic of vector rotation, the return to the past to design a new system. The logic of actualization is precisely the expression of this mechanism. The Japanese case should not be privileged as unique.[5] This is a typical example which shows a society can use the past-oriented vector to design a new social system. Second, the "rearrangement of things" that Maruyama mentioned may mean that there is a box hidden somewhere,[6] and that one can draw elements from it to remember. This way of looking at things tends to deny irreversible and continuous time. As Takamura's memory goes from his ancestors to Rodin, an element to be remembered is chosen depending on the situation in which the individual is involved.

Zero-vector situation

The memory focuses on the past-oriented vector even though afterwards the vector can turn to the future by vector rotation. It tends to deny the order of time lead by the future-oriented vector. From the moment when the past-oriented vector is socially recognized, the distinction between past, present, and future becomes less clear. This provokes the state of uncertainty in which one does not know any more where vectors go. This is precisely the zero-vector situation. Everything we have discussed contains elements of the zero-vector situation. Postmodern thinking starts from the moment when there are no more norms or orientation. Halbwachs assumes that collective memory is alive and changes according to the situation. It always has an element of uncertainty. Japanese thought has no historical axis, and a thought that is considered to have disappeared reappears.

The zero-vector situation is well described by Chômei Kamono, Japanese essayist of the twelfth and thirteenth centuries. He wrote it in his *Hôjôki, Visions of a Torn World*.

> On flows the river ceaselessly, nor does its water ever stay the same. The bubbles that float upon its pools now disappear, now form anew, but never long. And so it is with people in this world, and with their dwellings.
>
> In our dazzling capital the houses of high and low crowd the streets, a jostling throng of roof and tile, and have done so down the generations – yet ask if this is truly so and you discover that almost no house has been there from of old. Some burned down last year and this year were rebuilt. Others were once grand mansions, gone to ruin, where now small houses stand.
>
> (Kenhô and chômei, 2013:5)

He designates in his essay the ephemeral nature of human being and society. This conception of time and society is not a simple poetical one but results from the observation about his society. In particular, he deals with the calamities of society. His essay describes the great fire of 1177, the tornado of 1180, the famine of 1181 to 1182, and the great earthquake of 1185. Since society is constantly changing and it is difficult to keep a fortune for a long time, it makes no sense to keep property. Chômei notes the futility of the effort to build a new home since it would be destroyed for one reason or another (Kenkô and chômei, 2013:6).

This period was marked not only by these disasters but also by great social change. Indeed, already in 1180, when a rising samurai clan Taira practically took power although officially, the power was still in the hands of the emperor and the aristocracy. The Taira clan tried to transfer the capital from Kyoto to Kobe in 1180, but in the same year, Taira's rival clan, Minamoto, which had fought through Taira and exiled to the east, attacked the Taira clan. The war started between the two clans and ended with Minamoto's victory in 1185. The chief of the clan, Minamoto no Yoritomo transferred the capital to Kamakura far away from Kyoto in 1192 to cut off all the influences of the emperor and the aristocrats. It is from this date that the samurai class took power until 1867. Chômei didn't give any word about the war because he only wrote what he actually saw and observed as a reporter; the war happened outside Kyoto where he lived.

Chômei's idea is curiously close to the contemporary physicist Carlo Rovelli, who states that "the world is nothing but change" (Rovelli, 2018:85). He also makes the following remark.

> We can think of the world as made up of *things*. Of *substances*. Of *entities*. Of something that *is*. Or we can think of it as made up of events. Of *happenings*. Of *processes*. Of something that *occurs*. Something that does not last, and

that undergoes continual transformation, this is not permanent in time. The destruction of the notion of time in fundamental physics is the crumbling of the first of these two perspectives, not of the second. It is the realization of the ubiquity of impermanence, not of stasis in a motionless time.

(Rovelli, 2018:86–87)

Rovelli points out that thinking of a world as a collection of events is the only way compatible with relativity. There is only change and process. Substantial things no longer make sense, and only the event counts in the world of physics. This way of thinking that nowadays covers different fields is not peculiar to the "modern" consumer society. The thirteenth-century writer had the same way of thinking. Under what conditions, then, does this way of thinking emerge? The answer is as follows: it emerges when change is not only observed but considered as the starting point for understanding the world. It is no longer order, stability, and structure but change, mobility, and movement that count.

Identity of space

The Age of Preservation begins with the increased importance of the past-oriented vector while the future-oriented vector weakens. And then, little by little, the zero-vector situation prevails. There is no longer any sense of duration. This is the reason why there is no orientation toward the past or toward the future. The world is constantly changing, and only change matters. The society is fluid, and it does not stay in the same condition. Time has been a major source of power. But its strength is weakening as the zero-vector situation develops. If this is so, is the attempt to preserve finally in vain, while the Age of Preservation seems to preserve everything? The answer is no because space replaces time in a way for the preservation of heritage. It is then necessary to reconsider preservation of cultural heritage with reference to the space. In fact, society as well as the individual[7] cannot exist with this observation of entropy like a physicist; they need stability. This stability can be provided by space. Indeed, when we can see the same landscape every day, the space is stable. The stability of space is a condition *sine qua non* to show the continuity of the society. We call this situation identity of space. The identity of the space is assured, on the one hand, when, on the whole, the integrity of the district, or even of the city, is not called into question for one reason or another and, on the other hand, when the inhabitants recognize this integrity and actually experience it. And this integrity can be experienced and observed as a landscape. Indeed, individuals do not live independently of the landscape in which they live. Landscapes are produced by the community to which individuals cannot remain unrelated. Individuals have no choice but to live in the spaces produced by the community they belong to.

Japanese philosopher Tetsurô Watsuji highlights this relationship between body experience and space. Watsuji defines the space where an individual lives and understands himself/herself as *fûdo*. Watsuji discusses the self which exists in the *fûdo* by taking "cold" as an example. The cold and the self who feels it are not divided into the objective cold and the subjective self. "When we feel the cold, our selves already exist in the outside cold" (Watsuji, 1979:12). In addition, "the cold is experienced by 'us' and not just by 'me'" (Watsuji, 1979:13). Cold is felt within us or, in other words, represents the relations among us. We talk about the weather because we share the experience of feeling the cold (or heat). And, in turn, we feel that we stand on the same footing when we are in the *fûdo* of the locality, which may have strong mountain winds, *yama oroshi* or strong dry winds, *karakkaze* peculiar to the area. Watsuji describes this process as "self-understanding" of the cold in *fûdo*. In this process, how we subjectively feel the cold does not matter. Feeling the cold is a social process through which we take various measures, such as protection against the cold.

Since our body is always linked to the landscape, if there is a change to be made by work to a landscape, it must be announced in advance. Residents must be well informed to avoid creating a feeling of uncertainty. Even if fluidity is a reality, it should be concealed somehow. There are, however, circumstances in which the identity of the space can no longer be assured. This is the case of disaster. It brings a great change and daily life is interrupted. Disaster areas lost precisely the identity of space. It is therefore necessary to rebuild the disaster area and restore spatial order. If traces of catastrophe remain for a long time, the negative past they embody cannot be erased. If nothing is done, the uncertainty

Figure 6.1 The old shopping street in Takasago city
Source: Photo by the author.

is physically visualized. The future of the area is completely uncertain. This kind of space makes a disorder in the area. This situation, in which what should have passed away is still there, is defined as "frozen time–space". The past remains bare, exposed, or remaining intact in the present in this situation.

There are also cases where change is slower and harder to recognize in everyday life, but the situation of frozen time–space manifests itself and the identity of the space is lost. We take an example of an old port city, Takasago in Hyogo prefecture, Japan. Today, there is almost no trail of the port city. We just find a signboard that indicates the origin of the city. In this city, around the station, there is a shopping mall. In the same area, there are also factories. On the contrary, in the old shopping street (Figure 6.1), most shops are definitely closed except a sake dealer. The shopping street breaks the identity of the space of the area, because in this space, to some extent, the time is frozen. The shopping street is different from its surroundings. So, it brings a sort of crack in the area. The identity of space is lost.

Frozen time–space and violence

The disorder caused by the loss of identity of space manifests itself in the emergence of violence. Indeed, in Takasago, a series of violence occurred. In 1983, a 14-year-old boy attacked his classmate by a knife who had exercised bullying. There was a case of suicide in 2006; a 15-year-old boy killed himself to protest the mistreatment in the school he attended, etc. From our research on bullycide, suicide caused by bullying, we have found the relationship between bullycide and place where the bullycide cases occurred (Ogino, 2008b, 2012). In fact, a comparison of the population size of the communities with 133 bullycide cases that occurred from 1979 to 2009 shows that there is a relationship between population size and the frequency of bullycide. We calculated the ratio of bullycide cases to the population of the municipality where the suicide cases occurred, for the year during which they occurred. The ratio was converted to the number of cases per 100,000 persons. Based on this method of calculation, large cities with populations over 1,000,000 had 0.04 bullycide cases per 100,000 persons. Similarly, cities with populations between 100,000 and 1,000,000; those with population between 30,000 and 100,000; and those with population less than 30,000 had 0.30, 1.68, and 6.39 bullycide cases, respectively. This indicates that the smaller the population size, the higher the likelihood of bullycide cases occurring.

What, then, are the characteristics of the communities in which bullying victims actually committed suicide? It is virtually impossible to investigate each and every bullycide case. We thus focus on a portion of these suicide cases with a particular characteristic: cases which involve money. Some of the bullying cases that led to the victim's suicide involved activities similar to racketeering or blackmailing. A typical example of this is the bullycide case that occurred in December 1994. The victim, a 14-year-old boy, explains the reason for his suicide in his quite lengthy suicide note:

> Four persons (sorry I can't mention their names) always took money from me. Today, I just couldn't find money to bring to them. If I stay alive . . .

I killed myself because they took 40,000 yen from me today again. Even if I have no more money and say "I couldn't get any money", they would just bully me and tell me to try again.

(the evening edition of the Asahi Shimbun dated 5 December 1994)

This boy is attempting to explain the reason for his suicide in the context of money. This suggests that the bullies communicated mainly through the medium of money. Friendship does not, by its nature, involve money. Indeed, the general understanding of friendship is that it develops when the potential friends do away with cash exchanges between them. Based on the view that friendship is incompatible with money, no friendship seems to have existed between this boy and the bullying boys.

However, the bullying developed in a world where the generally accepted view – that friendship is "pure" and never involves money as a medium – does not hold. This suggests market economy has spread into the realm of friendship and a world is being formed where friendship is a commodity that can be purchased with money. Or, it may be more appropriate to say that the world of friendship is being rearranged to simulate the behavioral principles of a market economy.

Of the 133 cases of bullycide identified by us, 29 clearly involved money, as evidenced by the suicide note and/or other evidence. We have investigated all these communities. Results of the investigation indicate many similarities among these communities. First, the population changes during the two decades immediately preceding the bullycide cases indicate a population increase in almost all of the communities. The construction of industrial and housing complexes had produced rapid population growth in these locales. But the most important result of our research is that these communities all show the same landscape characteristics. We have found the following common components in the scenes of the communities where the bullying victims killed themselves:

1   Mountains and forests, rivers, lakes and/or beaches: These landforms constitute the natural environment but, needless to say, this landscape has been used and modified by people.
2   Farmland.
3   Tile-roofed, wooden houses.
4   Old shopping arcades: These were often built before the war and are often in decline due to a lack of successors.
5   Shrines, Buddhist temples, and graveyards: These communities still have shrines and Buddhist temples, which used to occupy an important place in the community, as well as graveyards, where community members worship their ancestors.
6   Public facilities: Public facilities, such as public offices and educational institutions, are located in the central district of the community. They are reinforced concrete buildings.
7   Housing developments: After the foundation of the Nihon Jûtaku Kôdan, Urban Development Corporation, in 1955, an increasing number of housing complexes, made of reinforced concrete, were constructed. The number of

rental houses constructed by the corporation peaked in 1971. The communities are dotted with a variety of houses, ranging from relatively old housing complexes constructed in the 1960s to privately owned condominium buildings only recently completed.

8   Factories and industrial complexes: These, too, range from independent factories to industrial complexes comprising a number of companies.

9   Roadside shops: The communities have large suburban shopping facilities, such as shopping malls, supermarkets, game arcades, fast-food restaurants, family restaurants, discount electrical appliance stores, and karaoke rooms.

These components of the scenes are defined as spatial symbols. The spatial symbols can be broadly divided into two categories. The first category comprises symbols made mainly of natural materials. The natural environment, farmland, wooden houses, and shopping arcade buildings constructed before World War II fit into this category. The second category comprises symbols made mainly of synthetic materials.

Synthetic materials became widely used after the 1970s. They are now indispensable in our everyday life. Rivers are lined with concrete embankments. Even in wooden houses, synthetic materials are used in the kitchen, bathroom, and toilet facilities. Still, while the spatial symbols listed from (1) through (5) are constructed based on the principle that the primary materials must be natural, underlying the construction of those listed from (6) through (9) is a preference for synthetic materials.

In the communities where the bullying victims killed themselves, many different spatial symbols coexist in small areas. These areas also have symbols of both the first and second categories. For example, we may find housing or industrial complexes dotted with farmland. This type of landscape is defined as a "destructured landscape". The greater the variety of spatial symbols found in a scene, the more destructured it becomes. A combination of spatial symbols of the first and second categories makes the landscape particularly destructured.

In addition to the spatial symbols listed earlier from (1) through (9), there is another spatial symbol that is an essential component of destructured landscapes, namely, vacant lots. Vacant lots are regarded as the tenth spatial symbol constituting destructured landscapes. Destructured landscapes have undeveloped, vacant lots. Weeds have colonized these lots. They are "blank" spaces in the true sense of the word, in that no one knows for sure how they will be used. These blank spaces bring heterogeneity to the entire scene. This blank space is precisely a frozen time–space. It is an abandoned space, and it is no longer an object of exploitation. By adding to the empty space, a street where most of the stores are closed or a house that has not been inhabited for a long time, etc., constitute the frozen time–space. If the landscape is destructured, the space left behind by change is more likely to exist. At the same time, the feeling of uncertainty expands.

In contrast to the ten spatial symbols that are components of destructured landscape, there are certain spatial symbols that are incompatible with a destructured landscape. One such symbol is skyscrapers. No skyscrapers are found in destructured landscapes. Well-maintained parks are another component less often found in destructured landscapes. Yet another symbol incompatible with destructured landscapes is terminals. Train or bus terminals do not exist in destructured landscapes. Most communities with destructured landscapes are distant from bus stops or train stations. Skyscrapers and transportation terminals are spatial symbols that are characteristic of large cities. These components contain "transparent spaces" characterized by sufficiency, odorlessness, and safety. Sufficiency means that people can obtain all the necessities of life. One never needs to starve in a big city. Shop windows symbolize material sufficiency. In former times, fish shops and greengroceries had strong smells of the products they handled, due to the nature of these products. In supermarkets, food products are wrapped in plastic packages, which shut off the products' natural smells. Odorlessness is the first requirement to show that the store has been disinfected and cleaned. The last point about transparent spaces is the pursuit of complete safety, with surveillance cameras positioned everywhere. Some of these spaces look as though they are healing spaces, giving the impression of utopia. Transparent spaces do not have a destructured nature, because they are dominated by synthetic materials.

Henri Lefebvre describes the process through which a city incorporated rural villages in Western Europe during the sixteenth century. Lefebvre's understanding is that a city continuously "abstracts" the "natural space" of rural villages. He describes the process of containment of "peculiarity" by a "universal existence" as the expansion of "abstract space". This applies in part to the process of the production of destructured landscapes. This is a process through which the peculiarity of *fūdo*, as defined by Watsuji, is lost, and scenes with combinations of different spatial symbols expand. Phenomena that characterize the *fūdo*, such as *karakkaze* or *yama oroshi*, become less and less significant. Once the roads have been paved, people no longer have to worry about dust. Instead, exhaust gas from cars will pollute the area, and the community will face new problems, such as environmental issues. These problems are no longer caused by the *fūdo*, because exhaust gas emissions do not stem from any specific *fūdo*. Exhaust gas emissions could occur anywhere with increased automobile traffic. Exhaust gas is a product of chemical reactions of the fuel. It is a product that symbolizes the dominance of chemistry. This is the dissolution process of what Watsuji describes as the historic *fūdo*. These scenes form spaces are abstract and lack singularity.

According to Lefebvre, when a natural space has lost significance and the abstraction has reached an advanced stage, "spatial codes" come into existence. Spatial codes comprise: (1) alphabets and vocabulary; (2) grammar and rhetoric; and (3) stylistics, each of which pertains to the components of space (Lefebvre, 2000:312). Item (1) signifies a list of the components of space, ranging from water and air to bricks and concrete blocks, and of materials and tools necessary for the construction of the space. Item (2) indicates the arrangement of individual spatial components in a manner that ensures unity of the whole. Item (3) presents a solution on an aesthetic level which describes the harmony, order, and products of the space. In short, space is encoded and treated as codes, like language. Just as language has grammar and beautiful styles, space must be ordered and arranged beautifully. Space must maintain unity and order and must be constructed systematically and presented intentionally. The encoding of space, as referred to by Lefebvre, has been carried out to perfection in cities, which formed the focus of Lefebvre's attention. This was particularly the case in large modern cities. In contrast, destructured landscapes occur in communities which have not been fully urbanized and whose space has thus not been fully encoded. Lefebvre did not take into account this space, which is neither natural nor abstract.

Destructured landscapes are likely to appear during a period when the community's main activity shifts from farming to factory work and, further, the consumption lifestyle as symbolized by roadside shops is introduced. These changes occur based on the comprehensive development program. Comprehensive development refers to the process of encoding a space. Once a space is produced based on a certain code, the space can no longer be analyzed other than through examining how the various symbols are arranged based on the code used. However, a destructured landscape is not completely controlled by the encoding of the space because, if so, then the landscape would not have become destructured. A destructured landscape is a product of incomplete encoding. Because of this, there is no more common space that the inhabitants can share. A destructured landscape separates, thus, the human body from the space and promotes isolation of the body. Especially the children who are brought up with the destructured landscape cannot appropriate an identity of space. This isolation makes it difficult for some of them to communicate with others and might drive some to suicide.

Frozen time–space and identity of community

So far, it has been shown that social order can be understood in terms of time and space. But, to ask if cultural heritage plays a certain role in society, it is necessary to consider the relationship between time and space. Giddens has precisely noted that time, space, and repetition are intertwined (Giddens, 1979:204). The concept of *Fúdo* illustrates well this relationship. For example, *karakkaze* is the wind that blows in early spring in the northern region of Japan. Every year, this strong wind returns and announces the arrival of spring. However, it is not always a beneficial phenomenon. The fields were often damaged by *karakkaze*, and measures had to be taken against it. The natural and climatic phenomenon specific to the region, the shared cyclic chronology, that is, the annual arrival of *karakkaze*, and the space that the inhabitants live in on a daily basis are closely linked. The chronology of a community corresponds to its environment. The temporal order is represented spatially from the sowing to the harvest, for example, for the farmers. However, the separation of time from space (Giddens, 1990:19) took place with the development of capitalism. The notion of time and space each acquires an autonomy. The two have an order independent of each other. The timetable is a temporal management device used in the factory and in the school. Students and workers are expected to respect it. That said, as Giddens remarked, the timetable became "a time-space device indicating both when and where" (Giddens, 1990:20). It means that regardless of the type of society, there exists a correspondence between temporal and spatial orders.

The destructured landscape loses this correspondence. In the destructured landscape, spatial symbols produced at different times exist without order and only represent two different categories of spatial symbols. Natural materials give us an impression of relative oldness and synthetic materials, novelty. The two categories only represent a dichotomous qualitative difference between the old and the new. The past and the present coexist without a meaning being given to this coexistence. In this circumstance, the region lacks the perspective for the direction it should take. Especially if a place cannot keep up with the change that the rest of the area experiences, its marked difference constitutes a frozen time–space. The old shopping street, in which most of the stores are closed, could not adapt to the new type of consumption. Such a place overtaken by time or a place that keeps its cursed past arouses uncertainty and must be erased from the present landscape to restore the correspondence between time and space.

As for the case of regions stricken by a natural or social disaster, if the reconstruction does not proceed and the damaged area is left in place for a long time, it means that what should have passed away is still there. Then, frozen time–space may appear. In the case of the little frequented shopping street, the contrast between the frozen space and the rest is marked, whereas in the case of the disaster area, the difference between the before and after disaster is explicit. But in both cases, the change that has occurred is experienced as exceptional, and society does not have enough means to dispel the uncertainty created. If the frozen time–space risks creating disorder and possibly giving rise to violence, it must, in

Figure 6.2 A street in Tatsuno city designated as Important Preservation District of Traditional Buildings

Source: Photo by the author.

any case, be dealt with. If nothing is done, the future of the area is completely uncertain. How, then, can the frozen time–space be dispelled?

To answer this question, it is necessary to give to the frozen time–space a social sense. Its reconstruction is indispensable for this purpose. The idea to make of the frozen time–space a cultural heritage appears precisely at this moment. For example, in the case of the shopping street, the first solution is to renovate the shopping street. If the little frequented shopping street was completely innovated and took back the prosperity of former times, it would be ideal. But usually there is a more modernized shopping mall next door that people prefer. This is how to make the shopping street a historical monument. This is what is happening in some districts. This is precisely the case of Tatsuno city in Hyogo prefecture of which Takasago City is also a part. A street was designated as the Important Preservation District of Traditional Buildings by the Agency for Cultural Affairs (Figure 6.2). In some cases, a district in whole becomes an ecomuseum. The process is the same for the disaster area. Two projects can be started at the same time just after the great disaster. The first project is the reconstruction of the disaster area. It aims to achieve its restoration as quickly as possible. The second project, which develops in the Age of Preservation, is the conservation of certain objects as memorials. In doing so, the cursed memory is socially exorcised. The cursed event is arranged in the time–space order. The cultural heritage institutions thus constitute an apparatus for recovering from the damage caused by the natural or social disaster.

Conclusion

We finally arrive at the conclusion of this book: if there is reason to be interested in cultural heritage, the spatial reconstruction using cultural heritage could resolve the state of frozen time–space by giving a new time–space order. We can even say that in fact, from the birth of the constitution of the cultural heritage, its goal was to dissipate the frozen time–space. This was the case for the creation of the Louvre just after the Revolution, which brought a great change. It should create a lot of frozen time–space because of the struggle between different parties. Another example is the aesthetic gaze that appeared in England in the eighteenth century (Woodward, 2001). Since that time, the ancient ruin was an object of painting. The beauty of ruin was discovered. Finding the beauty in the Roman ruin was the attempt to symbolically transform the frozen time–space into cultural heritage. Thus, the various Roman ruins have acquired historical and cultural value today and are included in the World Heritage List.

However, this role of cultural heritage institutions does not always work and meets with embarrassment. First, World Heritage listing creates a new contrast between the city with a World Heritage site and the city nearby. This is the contrast between Himeji and Takasago, which is next to Himeji. The Himeji castle (Figure 6.3) was inscribed on the World Heritage List in 1993. From Himeji station, there is the main street straight ahead, which leads to the castle. On both sides of the street, there are stores and offices that make up a lively district. In addition, there are tourists who visit the castle. On the other hand, in Takasago,

Figure 6.3 Himeji Castle and main street in Himeji city
Source: Photo by the author.

the city next to Himeji, as mentioned earlier, there is a frozen time–space, and this contrast between Himeji and Takasago marks the people of Takasago. An destructured landscape is created between Himeji and Takasago. The same situation can be observed in Nancy in France. Nancy is a central city of Lorraine, and its Place Stanislas, Place de la Carrière, and Place d'Alliance were inscribed on the World Heritage List in 1983. But as we move away from the city center, the depopulated landscape appears. A neighboring village teacher we interviewed reported that the bullying was frequent at school.

Second, World Heritage listing can cause serious conflicts. As we noticed, in the case of the Hashima Coal Mine, from its inscription on the World Heritage List, the struggle between Japanese nostalgic nationalism and the negative memory of the Korean former forced laborers began. Third, with regards to the negative heritage, it is necessary to take into account a gap between the physical reconstruction of the city, fixation of the representation on the event in question, and the mental recovery of the victims. In the case of Warsaw, the Old Town was perfectly reconstructed like before the destruction by Nazi troops in 1971. But only in 2004, a long time after the reconstruction of the Old Town, was the Warsaw Rising Museum founded to memorialize the Warsaw Uprising. This museum can be considered an official representation of the event. Another example is the Berlin Wall. When East and West Germany were unified, the first thing torn down was the Berlin Wall. Just a part of the wall is kept in the National Museum of

History and Topography of Terror still in 1998.[8] The terrain was not well organized, and student volunteers insisted on the need to keep walls so as not to forget the importance of the event and emotions that this event raised. It was only in 2010 that the new Topography of Terror Documentation Center and the newly designed terrain opened (Topographie des terrors, 2020).

It takes time until the conservation project is completed, because the conservation of a monument that recalls a negative event provokes controversy and the conservation project is difficult to achieve. The recovery of the city, the preservation of the monument, and the representation of the event do not completely erase the negative memories. The institutions of cultural heritage do not give the perfect remedy to get out of the situation in which the negative memory possesses them. Under what conditions then does the inscription on the World Heritage List or the designation of a site as cultural heritage become legitimate? In our opinion, it should not be a cultural nationalism or regionalism. This implies a closed and exclusive attitude. Given the negation of cultural nationalism, two conditions will then be required. First, whatever institutional frameworks are served, the inhabitants must take the initiative for the designation of a site as cultural heritage. Second, not only are places of exchange between inhabitants and visitors like the museum prepared but the exchange must substantially take place. To do this, the value that inhabitants give to both tangible and intangible heritage must be shared with visitors. And the ethics and logic of the exchange must not be imposed in advance by governments or international institutions. They must be created through a multitude of interactions. These interactions should constitute a gift system. The inhabitants and visitors will give, receive, and return like the gift cycle formulated by Mauss. A World Heritage site should not be a tourist place to be consumed like a consumer object but a gift object that the inhabitants give to the visitors, and the visitors who received it will give back one day in one way or another. Unlike the Maussian gift cycle, visitors do not need to give back directly to the inhabitants of the site, but, for example, they can give information about the site to those who have not yet visited it. Cultural heritage is the gift par excellence.

Notes

1 This question brings to ask whether government-led cultural heritage designation leads to "the death of culture".
2 This explanation is given by The Royal Institute of British Architects (The Royal Institute of British Architects, 2020).
3 The pertinence of this term should be examined. Is it "late" for what kind of people? For African people, their modernity is late? If so, it's late in comparison with what? In comparison with colonial period? Many countries have to be late modern before becoming modern.
4 This poem is cited by Maruyama (Maruyama, 1983:13).
5 In fact, there was the constitution of national history as Margaret Mehl has shown (Mehl, 1998). Simply, in the case of Japan, the past-oriented vector prevailed over the future-oriented vector.

6 This idea of a box hidden somewhere leads to consider modernity as a spatial notion. The notion of progress is not necessarily temporal because it can designate a social state superior to others. It presupposes the comparison of one society with another, but this comparison does not need to be situated in time. Indeed, in Japan, modernity no longer means the feeling of living in history but is first of all conceived as something external because it is brought from outside. Modernity is thus transformed into a spatial notion, and somewhere embodies modernity. There are two opposite types of discourse on modernity in the intellectual history of Japan: one is positive toward modernity, the other negative. But both have in common the conception of modernity as a spatial notion.

7 The observation of the zero vector does not prevent the author, like Chômei, from looking at the past. Indeed, the fact that he writes what he has seen already shows the interest in the past and, above all, the will to preserve his memory. The observation of the zero vector thus implies the will governed by the vector oriented toward the past and the vector oriented toward the future.

8 We conducted a research in October 1998.

References

Abbaye du Mont-Saint-Michel. 2020. Homepage. [online] available from: www.abbaye-mont-saint-michel.fr/en/Explore/L-histoire-de-l-abbaye-du-Mont-Saint-Michel#, last accessed 21 December 2020.

Abe, J. 2011. Doresuden ni okeru sekai isantôroku no tôrokumasshô ni itaru keika-keikanmondai no haigo ni arumono. *Nihon kenchiku gakkai koen bengaishû* 2011: 413–414.

Adam, B. 1990. *Time and Social Theory*. Philadelphia: Temple University Press.

Aizawa, S. 1973. Shinron. In *Nihon shisô taikei53 Mitogaku*. Tokyo: Iwanami.

Alele Museum. 2009. *Bikini Atoll, Nomination by the Republic of the Marshall Islands for Inscription on the World Heritage List 2010*. Homepage. [online] available from: http://whc.unesco.org/uploads/nominations/1339.pdf, last accessed 20 December 2020.

Althusser, L. 1976. *Positions (1964–1975)*. Paris: Les Éditions sociales.

Anderson, B. 1991. *Imagined Communities: Reflections on the Origin and Spread of Nationalism*. New York: Verso.

Ariès, P. 1975. *Essai sur la mort en Occident: du Moyen-Âge à nos jours*. Paris: Seuil.

Askew, M. 2010. The Magic List of Global Status, UNESCO, World Heritage and the Agendas of States. In Labadi, S. and Long, C. (Eds.), *Heritage and Globalisation*. Abingdon and New York: Routledge.

Bandarin, F. 2008. Forward. In Martin, O. and Piatti, G. (Eds.), *World Heritage and Buffer Zones Patrimoine mondial et zones tampons*. Paris: UNESCO World Heritage Center.

Baudrillard, J. 1968. *Le Système des objets*. Paris: Denoël.

Baudrillard, J. 1970. *La Société de consummation*. Paris: Gallimard.

Bauman, Z. 2000. *Liquid Modernity*. Cambridge: Polity Press.

Bauman, Z. 2005. *Liquid Life*. Cambridge: Polity Press.

BBC. 2020. Nepal's Kathmandu Valley Treasures: Before and After, Homepage. [online] available from: www.bbc.com/news/world-asia-32472307, last accessed 20 December 2020.

Benjamin, W. 2005. Homepage. [online] available from: www.marxists.org/reference/subject/philosophy/works/ge/benjamin.htm, last accessed 13 December 2020.

Bourdieu, P. 1979. *La Distinction. Critique sociale du jugement*. Paris: Editions de Minuit.

Bourdieu, P. 2002. *Le Bal célibataire: Crise de la société pysanne en Béarn*. Paris: Seuil.

Breen, J. 2015. *Shinto monogatari*. Tokyo: Yoshikawa kobunkan.

Cameron, C. 2008. From Warsaw to Mostar: The World Heritage Committee and Authenticity. *The Journal of Preservation Technology* 39, 2/3: 19–24.

Caruth, C. (Ed.). 1995. *Trauma: Explorations in Memory*. Baltimore: Johns Hopkins University Press.

Chiva, I. and Lévi-Strauss, C. 1992. Qu'est-ce qu'un musée des arts des traditions populaires? *Le Débat* 70: 165–173.

Connerton, P. 1989. *How Societies Remember*. Cambridge: Cambridge University Press.

Dan, I. 1928. Honpô hakubutsukan ni kansuru shomondai. *Hakubutsukan kenkyû* 1–3: 3–4.

Decaëns, H. 2013. *Mont-Saint-Michel d'antan*. Paris: Éditions Hervé Chopin.

Descartes, R. 1966. *De la méthode*. Paris: G.-F. Flammarion.

Dôsako, S. 2009. Hagi hansharo kanrensiryô no chôsahôkoku (dai ippô). *Hagi Hakubutsukan Hôkokusho* 5: 1–5.

Drozdzewski, D., De Nardi, S., and Waterton, E. (Eds.). 2016. *Memory, Place and Identity: Commemoration and Remembrance of War and Conflict*. Abingdon and New York: Routledge.

Duncan, C. 1995. *Civilizing Rituals inside public art museums*. Abingdon and New York: Routledge.

Durkheim, E. 1912. *Les Formes élémentaires de la vie religieuse*. Paris: PUF.

Durkheim, E. 1978. *De la division du travail social*. Paris: PUF.

Ejima, S. 1977. Hibakukeiken ni kansuru shinborizumu no bunseki. *Shôgyô Keizai kenkyûjohô* 13: 115–139.

Eliade, M. 1969. *Le Mythe de l'éternel retour*. Paris: Gallimard.

Franklin, A. 2019. *Anti-Museum*. Abingdon and New York: Routledge.

Giddens, A. 1979. *Central Problems in Social Theory: Action, Structure and Contradiction in Social Analysis*. Berkeley and Los Angeles: University of California Press.

Giddens, A. 1990. *The Consequences of Modernity*. Stanford: Standard University Press.

Greenwood, D. 1989. Culture by the Pound: An Anthropological Perspective on Tourism as Cultural Commoditization. In Smith, V.L. (Ed.), *Hosts and Guests the Anthropology of Tourism*. Philadelphia: University of Pennsylvania Press.

Guibal, J. 1992. Quel avenir pour le musée des Arts et des Traditions Populaire? *Le Débat* 70: 157–164.

Hagi city. 2020. Homepage. [online] available from: www.hagishi.com/en/world-heritage/, last accessed 17 December 2020.

Halbwachs, M. 1925. *Les Cadres sociaux de la mémoire*. Paris: PUF.

Halbwachs, M. 1941. *La Topologie légendaire des évangiles en Terre Sainte*. Paris: PEUF.

Hamada, T. 2014. Sensô isan no hozon to heiwa kûkan no seisan. *Rekisi hyôron* 772: 20–34.

Harvey, D. 1989. *The Condition of Postmodernity*. Oxford: Basil Blackwell.

Hobbes, T. 1651. Leviathan, Homepage. [online] available from: www.gutenberg.org/files/3207/3207-h/3207-h.htm#link2H_4_0005, last accessed 13 December 2020.

ICOMOS. 2020. The Nara Document on Authenticity, Homepage. [online] available from: www.icomos.org/charters/nara-e.pdf, last accessed 17 December 2020.

Inaga, S. 1983. La perspective linéaire au Japon et sonretour en France. *Acte de la recherche en sciences sociales* 49: 29–45.

Inoue, T. 1908. *Rinri to Kyôiku*. Tokyo: Kôdôkan.

Jeudy, H.-P. 1986. *Les Mémoires du social*. Paris: PUF.

Jidai eno messêji (Ed.). 1997. *Genbaku dômu sekaiisanka eno michi*. Hiroshima: Genbaku dômu no sekaiisanka o susumeru kai.

Kamioka, N. 1987. *Nihon no kôgaishi*. Tokyo: Sekai Shoin.

Kenkô and çhômei. 2013. *Essays in Idleness and Hôjôki*. London: Penguin Books.

Kimura, S. 2014. *Sangyô isan no kioku to hyôshô*. Kyoto: Kyoto daigaku gakujustu shuppankai.

Kishimoto, N. 2017. Furuichi Mozu kohungun no hisôsha. In Imao, F. and Takagi, H. (Eds.), *Sekai isan to tennoryôkohun o tou*. Kyoto: Shibunkaku.

Kondô, S. 2014. *Fujisan sekai isan eno michi*. Tokyo: Mainichi Shinbunsha.

Labadi, S. and Colin, L. (Eds.). 2010. *Heritage and Globalisation*. Abingdon and New York: Routledge.

Lefebvre, H. 2000. *La Production de l'espace*. Paris: Anthropos.

Lévi-Strauss, C. 1962. *La Pensée sauvage*. Paris: Plon.

Lucky Dragon Peace Association (Ed.). 1986. *Haha to ko de miru daigo hukuryúmaru*. Tokyo: Sôdobunka.

Lyotard, J.-F. 1988. *Le Postmoderne expliqué aux enfants*. Paris: Galilée.

Maclellan, A. 1994. *Inventing the Louvre: Art, Politics and the Origins of the Modern Museum in Eighteenth-Century Paris*. Cambridge: Cambridge University Press.

Malreaux, A. 1930. *La Voie royale*. Paris: Grasset.

Manin, Pierre. 1994. *La Cité de l'homme*. Paris: Fayrard.

Maruyama, M. 1983. *Nihon no Shisô*. Tokyo: Iwanami.

Mauss, M. 1978. *Sociologie et anthropologie*. Paris: PUF.

Mehl, M. 1998. *History and the State in Nineteenth-Century Japan. Macmillan*. London: Palgrave Macmillan.

Misztal, B. 2003. *Theory of Social Remembering*. Berkshire: Open University Press.

Miyata, N. 1989. *Edo no chiisana kamigami*. Tokyo: Seidosha.

Mizuno, N. (Ed.). 1998. *Senjiki Shokuminchi tochi siryô*, Vol. 7. Tokyo: Kashiwa shobô.

Nagasaki zainichi chôsenjin no jinken o mamorukai (Ed.). 2016. *Gunkanjima ni mimi o sumaseba*. Tokyo: Shakai Hyoron sha.

Nanazawa, T. 2010. Doitsu · erubegawa ni okeru hashi no kensetsu to sekai isan taitoru masshô nitsuite no chôsa. *Kokudokôtsu seisaku kenkyú* 89: 1–50.

Naono, A. 2015. *Genbaku taiken to sengo nippon*. Tokyo: Iwanami.

National Institute of Tourism, Tourism Statistics Department. *Tourism Statistics Report*, April 2020. Homepage. [online] available from: https://seishiron.com/wp-content/uploads/2020/06/CAM042020.pdf, last accessed 13 December 2020.

Nietzche, F. 1994. *On the Genealogy of Morality*. Cambridge: Cambridge University Press.

Nobile, P. (Ed.). 1995. *Judgement at the Smithsonian*. New York: Marlowe & Company.

Nojima dansô hozon Hokudan kinen kôen. 2020. Homepage. [online] available from: www.nojima-danso.co.jp/nojima.html, last accessed 20 December 2020.

Nora, P. 1984. *Les Lieux de mémoire 1 la République*. Paris: Gallimard.

Ogino, M. 1995. Logique d'actualisation. Patrimoine et le Japon. *Ethnologie française* XXV: 57–64.

Ogino, M. 1998. *Shihonshugi to tasha*. Nishinomiya: Kwansei gakuin daigaku shuppankai.

Ogino, M. 2002. *Bunka isan no shakaigaku*. Tokyo: Shinyôsha.

Ogino, M. 2008a. *Scams and Sweeteners. Sociology of Fraud*. Melbourne: Trans Pacific Press.

Ogino, M. 2008b. Hybrid Landscapes: Where Children are bullied to Death. In Kosaka, K. and Ogino, M. (Eds.), *A Quest for Alternative Sociology*. Melbourne: Trans Pacific Press.

Ogino, M. 2012. *Kaihatsu kúkan no bôryoku*. Tokyo: Shinyôsha.

Ogino, M. 2013. Structuralism, Post-Structuralism, Thereafter. In Elliott, A., Katagiri, M. and Sawai, A. (Eds.), *Japanese Social Theory. From Individualization to Globalization in Japan Today*. Abingdon and New York: Routledge.

Ogino, M. 2015. Sociology of Collective Memory. In Wright, J. (Ed.), *Encyclopedia in Social and Behavioral Sciences*, Vol. 4. Oxford: Elsevier.

Ogino, M. 2016. Undercurrents in Japanese Cultural Heritage Management. In Matsuda, A. and Mengoni, L.-E. (Eds.), *Reconsidering Cultural Heritage in Asia*. London: Ubiquity Press.

Olick, J. 2007. *The Politics of Regret*. Abingdon and New York: Routledge.

Poulot, D. 1997. *Musée nation patrimoine 1789–1815*. Paris: Gallimard.

Raffles, S. 1988. *The History of Java: Complete Text*. Oxford: Oxford University Press.

Ringbeck, B. and Rössler, M. 2011. Between International Obligations and Local Politics: The Case of the Dresden Elbe Valley Under the 1972 World Heritage Convention. *Informationen zur Raumentwicklung* 3, 4: 205–211.

Rovelli, C. 2018. *The Order of Time*. London: Penguin Books.

The Royal Institute of British Architects. 2020. Homepage. [online] available from: www.architecture.com/explore-architecture/postmodernism, last accessed 20 December 2020.

Said, E. 2003. *Orientalism*. London: Penguin Books.

Sakamoto, D. and Goto, K. 2005. *Gunkanjima no isan*. Nagasaki: Nagasaki shinbunsha.

Satô, N. 1977. Keizai yôryaku. In *Nihon shisô taikei 45 Andô Shôeki Satô Nobuhiro*. Tokyo: Iwanami.

Sloterdijk, P. 2009. *Terror from the Air*. Los Angeles: Semiotext.

Takagi, H. 2002. Kindai tennosei to kodai bunka. In Amino, Y., Kabayama, K., Miyata, N., Yasumaru, Y., and Yamamoto, K. (Eds.), *Tenno to ôken o kangaeru 5 Ôken to girei*. Tokyo: Iwanami.

Takahashi, S., et al. 1955. Jûyô mukei bunkazai geinô o megutte. *Geijutsu shinchô*, July: 281–295.

Takahashi, T. 1995. Kiokusarenumono katarienumono. In Ukai, S. and Takahashi, T. (Eds.). *Shoah no shôgeki*. Tokyo: Miraisha.

Takamine, S. 2015. *Mahsharu shotô owarinaki kakuhigai o ikiru*. Tokyo: Shinsensha.

Teeuwen, M. and Breen, J. 2017. *A Social History of the Ise Shrines*. London: Bloomsbury.

Tochigiken Nikkôshi. 2020. Ashio Dôzan Nihon no kindaika sangyôka to kôgai taisaku no kiten, Homepage. [online] available from: www.city.nikko.lg.jp/bunkazai/gyousei/shisei/sekaiisan/documents/ashiodouzan-teiansho.pdf, last accessed 19 December 2020.

Tokyo Kokuritsu Hakubutukan (Ed.). 1973. *Tokyo kokuritsu hakubutukanshi*. Tokyo: Daiichi hôki.

Topographie des terrors. 2020. Homepage. [online] available from: www.topographie.de/en/the-historic-site/documentation-center/, last accessed 14 December 2020.

Umehara, T., et al. (Eds.). 1996. *Jômon bunmei no hakken*. Tokyo: PHP kenkyusho.

UNESCO. 2013. Operational Guidelines for the Implementation of the World Heritage Convention, Homepage. [online] available from: http://whc.unesco.org/archive/opguide13-en.pdf, last accessed 17 December 2020.

UNESCO. 2015. *World Heritage 39 COM*. Paris: UNESCO.

UNESCO. 2020a. Homepage. [online] available from: https://unesdoc.unesco.org/ark:/48223/pf0000082464.page=66, last accessed 17 December 2020.

UNESCO. 2020b. Homepage. [online] available from: http://whc.unesco.org/en/conventiontext/, last accessed 14 December 2020.

UNESCO. 2020c. Homepage. [online] available from: https://whc.unesco.org/en/list/95/, last accessed 14 December 2020.

UNESCO. 2020d. Homepage. [online] available from: https://whc.unesco.org/en/list/31/, last accessed 14 December 2020.

UNESCO. 2020e. Homepage. [online] available from: https://whc.unesco.org/en/list/775/, last accessed 14 December 2020.

UNESCO. 2020f. Homepage. [online] available from: https://whc.unesco.org/en/list/1449, last accessed 17 December 2020.

UNESCO. 2020g. Tomioka Silk Mill and Related Sites, World Heritage Nomination, Homepage. [online] available from: https://whc.unesco.org/uploads/nominations/1449.pdf, last accessed 17 December 2020.

UNESCO. 2020h. Homepage. [online] available from: https://whc.unesco.org/en/list/1484/, last accessed 17 December 2020.

UNESCO. 2020i. Homepage. [online] available from: http://whc.unesco.org/archive/serial-noms.htm, last accessed 17 December 2020.

UNESCO. 2020j. Homepage. [online] available from: https://whc.unesco.org/archive/2011/whc11-35com-INF9Ae.pdf, last accessed 17 December 2020.

UNESCO. 2020k. Homepage. [online] available from: https://whc.unesco.org/en/list/1418/, last accessed 19 December 2020.

UNESCO. 2020l. Homepage. [online] available from: https://whc.unesco.org/en/list/872/, last accessed 20 December 2020.

UNESCO. 2020m. Homepage. [online] available from: https://whc.unesco.org/en/list/600/, last accessed 20 December 2020.

UNESCO. 2020n. Homepage. [online] available from: https://whc.unesco.org/en/list/37/, last accessed 20 December 2020.

UNESCO. 2020o. Homepage. [online] available from: https://whc.unesco.org/uploads/nominations/1156.pdf, last accessed 21 December 2020.

UNESCO. 2020p. Homepage. [online] available from: http://whc.unesco.org/en/list/668, last accessed 21 December 2020.

UNESCO. 2020q. Homepage. [online] available from: http://whc.unesco.org/en/list/811, last accessed 21 December 2020.

UNESCO. 2020r. Homepage. [online] available from: https://whc.unesco.org/en/news/1267/, last accessed 21 December 2020.

UNESCO. 2020s. Homepage. [online] available from: https://whc.unesco.org/en/list/121/, last accessed 20 December 2020.

UNESCO. 2020t. Operational Guidelines for the Implementation of the World Heritage Convention. Homepage. [online] available from: http://whc.unesco.org/archive/opguide13-en.pdf, last accessed 17 December 2020.

Veyne, P. 2015. *Palmyre. L'irremplaçable trésor*. Paris: Albin Michel.

Watsuji, T. 1979. *Fúdo*. Tokyo: Iwanami Shoten.

Williams, R. 1960. *Culture and Society 1780–1950*. New York: Anchor Books.

Wolska, J. 2013. Rebuilding Warsaw's History. *The Warsaw Voice*, November 28.

Woodward, C. 2001. *In Ruins*. New York: Pantheon Books.

Yanagi, M., Koyama, F., et al. 1955. Jûyô mukei bunkazai (kôgei) o megutte. *Geijutsu shinchô*, August: 157–72.

Yanagita, K. 1989. *Tono monogatari*. Tokyo: Kadokawa.

Yanagita, K. 1990. *Yanagita Kunio Zenshû 28*. Tokyo: Kadokawa.

Yanagita, K. 2008. *The Legends of Tono*. Lexington: Lexington books.

Yashiro, Y., Oshima, G., et al. 1929. Bijutsu rekishi kôkohin hozon mondai ken-kyûkai. *Hakubutsukan kenkyu* 2–4: 7–11.

Yasukuni jinja (Ed.). 1996. *Sange no kokoro to chinkon no makoto*. Tokyo: Tentensha.

Yasumaru, Y. 1971. *Fuji-kô. In Nihon shisô taikei 67 minshû shukyô no shisô*. Tokyo: Iwanami.

Yukimura, M. 2017. Sekai isan tôroku kijun ni kansuru ichisiron: henkyô no sekai isan ni chakumokusite. *Kansai daigaku shakaigakubu kiyô* 48, 1: 91–111.

Zambelli, R. 1995. Le Bohneur pour tous ceux que j'aime. *Thérèse de Lisieux* 750: 1.

Index

798 Art District 107

A-bomb 29, 32–34, 37–39, 44, 46;
  survivors 32, 34, 37–39
absorption of the past into the present
  112
abstract space 133
Aceh Tsunami Museum 112
Active Fault 111–**112**
act of mourning 110
Act on Protection of Cultural Properties
  81
Adam, Barbara 46
aesthetic gaze 69–70, 74, 92n2, 136
Age of Preservation 3–6, 8, 29, 71,
  94–96, 98, 102–104, 107, 109, 111,
  113, 115–116, 118, 120, 128, 136
Akita school 67
Althusser, Louis 2
ambiguous existence 12–14
ancient burial mounds 64–65
Anderson, Benedict 3, 4, 16
Angkor 103–104; Wat 23–**24**
antisepsis 44, 60
APSARA 24–25, 117n5
archaeological site of Carthage 96
Ariès, Philippe 6
Art House Project 107
Ashio Copper Mine 90
Ashio Copper Refinery **91**
Auschwitz 121; -Birkenau 30, 32
authenticity 14, 48–49, 53, 67, 71n1

Banks of the Seine 96
Banteay Srey Community Tourism
  24–**25**, 47n1
Baudrillard, Jean 114, 117nn9–10
Bauman, Zygmunt 5, 114, 119–120
Benjamin, Walter 14
Bikini Atoll 23, 32, 44, **45**, 46

Bizen 86–88
Bonaparte, Napoleon 16
Bourdieu, Pierre 58, 85
bourgeoisie 14, 85
Brasilia 3
Breen, John 77–78
buffer zones 20, 26

Caen Memorial 41–42
capitalist desire 7, 13–15, 72, 104–105,
  116
capitalist system 13–15, 21n4
Carthage 96, 97–**99**
Caruth, Cathy 39
catastrophe 109, 119
collective memory 5, 39, 69, 92, 122,
  124–126
colonization 17, 25, 62, 74
conflict 124
consumer society 113
criterion for selection 40
cultural desire 116–117
cultural diversity 7, 70
cultural heritage 7–8, 13–17, 19–20,
  22–23, 26, 28–29, 34–35, 37, 44,
  46, 49, 51, 53, 57, 58, 62, 67, 69,
  70, 71–72, 74–75, 77, 81, 84–86,
  90–92, 94–97, 102–105, 107,
  109–111
cultural nationalism 62, 138
cultural ubiquity 117
culture of oblivion 116–117

Daisen kofun 65
D-Day Landing Beaches 41
destructured landscape 6, 132–135
diplomatic negotiation 66
diversity of world cultures 48, 50
double loan 119
doubling of the world 5, 26

Dresden Elbe Valley 100
Drozdzewski, Danielle 39
Dubrovnik 23; Old City of 27
Duncan, Carol 1, 9–10
Durkheim, Émile 10, 122–123

Ebisugahana Shipyard 53–**54**, 56
effects of antisepsis 44, 60
Egyptian expedition 16, 21n5
Eliade, Mircea 11, 110
El Jem 96
Enola Gay 37, 43
evolutionary time 120
evolution of the rating of the three
    artists 88
exhibition value 14, 21n4

fluidity 129
former Gorin Church 63, **64**
Fourvière 95
framework of memory 124
frozen time–space 130, 132, 136
fûdo 133
Fuji-kô 79
Fujisan site 67, 92n80
future-oriented vector 120

Genbaku Dome 3, 23, **30**, **31**–32,
    33–37, 44–46, 50, 57, 90, 119
Giddens, Anthony 7, 123
gift 10, 12; system 138
global capitalism 114
Godzilla 45
Gorin 63; Church 63, **64**
Grand Narrative 120

Hagi Castle Town 54, 56
Hagi Reverberatory Furnace 53
Hague Convention 7, 16, 19, 21n5
Halbwachs, Maurice 5–6, 123–124,
    126
Harvey, David 114, 119
Hashima Island **59**–60
hibakusha 32–33, 44
Hidden Christian Sites in the Nagasaki
    Region 51, 57, 62
Himeji castle 48, 136–**137**
Hiroshima Army Clothing Depot 37
Hiroshima Peace Memorial Museum 33,
    36, 40–42, 44
historical consciousness 94
Historic Centre of Warsaw 28–29, 119
history 120
Hobbes, Thomas 13, 18

Hukuromachi Elementary School 36
hypocenter 30, **36**

ICOMOS 28, 48, 57, 63, 66, 69–70,
    101
iconoclastic ideology 115
identity of community 134
identity of space 128–130
iemoto system 80
imperial ideology 65, 71n8, 72–74, 79,
    126
Implementation of the World Heritage
    Convention 15, 20, 28, 66
important intangible cultural property
    82
intangible cultural property 86
intangible links 69–70
integrity 7, 14–16, 44, 49, 53, 66, 67,
    103, 105, 107, 128
International World Heritage Expert
    Meeting on criterion 50
Ise Shrines 76–78, 83, 92
Iwami Ginzan Siver Mine 62

Japanese model of modernization 56
Jeudy, Henri-Pierre 6
Jingdezhen 107
Jomon Archeological Sites 51, 64, 65
Jomon culture 66

Kamono, Chômei 127
Kapal di atas rumah **113**
Kathmandu valley 109
Katsushika, Hokusai 67
Khmer Rouge 23–26, 103, 117n8
Killing Fields 26
Kokugaku 73–74, 80
Kondô, Seiichi 67, 69
Kôtarô Takamura 125

Labadi, Sophia 28, 48
landscape 128, 131
late modernity 123
Law for the Protection of Cultural
    Property 35, 57, 89
Lefebvre, Henri 20, 133–134
Legends of Tono 11
Lévi-Strauss, Claude 1, 95
lieux de mémoire 6
Lijiang 104–105, **106**–107, 117n6
linear conception of time 95
linear perspective 67
living culture 103
Living National Treasure 82

loan system 113, 119–120
logic of actualization 82
Louvre project 9
Lucky Dragon 32–33, 45
Lyon 95–**97**
Lyotard, Jean-François 120–122

Maclellan, Andrew 9
Manin, Pierre 120
Maruyama, Masao 125–126, 138n4
Matsudo Municipal Museum 3
Mauss, Marcel 10, 12, 138
Medina of Tunis 96
Meiji Industrial Revolution 51, 53–54, 57, 60, 66
Meiji Restoration 53–54, 60, 74–75, 80, 89
memory 124; of colonization 62
Memorial Cenotaph for the A-Bomb Victims 34
Miho no Matsubara 67, 69–**70**; pine tree grove 67
Mito school 72–74
Mitsubishi Shipyard 57, **58**
modernity 120–123, 138n3, 139n6; hidden side of 121; light 114
Mont-Saint-Michel 105–**106**; Abbey 105
Mount Fuji 67–**68**, 69–**70**, 78–79, 92n3
Mozu-Furuichi Kofun Group 64–65, 92
museological desire 1–2, 7, 9–10, 14–15, 17–18, 20, 29, 31, 59, 72, 74, 88, 90, 92, 95
museum 14, 74–75, 82–84, 88, 92, 95, 103, 111–112, 116

Naono, Akiko 38–39
Naoshima island 107
Nara Conference 7, 48–49
Nara Document 7, 28–29, 48–50, 70, 94, 116
narrative 7, 49, 51, 53, 56, 58–60, 62, 65–66, 92, 96
National Air and Space Museum 43
national history 72
nationalism 17
nation-state 1, 7–10, 13–14, 16–18, 22, 72, 77, 124–126
Naxi people 105
negative heritage 7, 23, 29, 42, 50, 137
new time–space order 136
Nojima Fault Memorial House 112

Nojima Fault Preservation Museum 111
Nora, Pierre 5, 122
nostalgia 118–119
nuclear colonialism 44–46

Ogino, Masahiro 82, 85, **88**, 122, 125, 130
Ohitayama Tatara Iron Works 54, 56
old Bridge Area of Mostar 29
old town of Warsaw 27–28
Omuta, Minoru 37
Operational Guidelines for the Implementation of the World Heritage Convention 15, 28, 66
Oradour sur Glane 29
order of recollection 10, 12–13, 34, 72, 74, 110, 116
orientalism 17
Orizuru No Kai 32–34
outstanding universal value 15, 41, 43, 45, 59, 62, 70, 95

Pacific War 32, 42, 65, 81, 125
pacifism 36–37, 42–44
Palmyra 115
Paris 1, 96
past-oriented vector 119, 123, 126
peace 22, 30–31, 34, 37, 42–44, 73
pilgrimage 78
politico-ritual institutions 75, 81, 84, 89, 92
postmodernism 121
post modernity 114, 122
principle of secrecy 83–84
progress 120
PTSD 122
public–private relationship 84

Raffles, Thomas Stanford 16
reflexive modernization 123
reflexivity 123
relative autonomy of the private sphere 84
representation of space 20
risk 120
Roman ruin 136
Rosetta Stone 16–17
Rovelli, Carlo 127–128

sacred center 11, 46, 110
Sannai-Maruyama ruins 66
Sasaki, Sadako **33**–34, 41
semantics of inscription 48, 51

Shôkasonjuku Academy 54, 56, 58
shopping street **129**, 130, 135–136
skyscraper 133
Sloterdijk, Peter 22, 29
space of consumption 104
spatial codes 133
spatial symbol 132
stability of space 128
surveillance 120
symbolism of the center 110
synthetic material 133
system of signs 114

Tajima Yahei's House **52**
Takasago 130–131, 136–137
Tatsuno **135**
Tetsuro Watsuji 128
Thirty-six Views of Mount Fuji 67
timetable 134
Tôdaiji 75–**76**
Topography of Terror 138;
    Documentation Center 138
tourism 104
tourism industry 104
traboules 95
transcendental value system 71
transparent spaces 133
transportation terminal 133
trauma 118, 122
traumatic situation 39

uncertainty 120
UNESCO 22, 48, 53, 94, 96, 98,
    100–105, 110–111
unique genealogy of the imperial family
    73

vector 118
vector rotation 119
Veyne, Paul 115

Waldschlößchenbrücke 100, 101, **102**
War 19, 22, 40, 42–44, 46, 50,
    126–127; Bosnian 29; heritage
    29–31
Warsaw Rising Museum 137
water-supply system 104
World Heritage Convention 15, 19–20,
    22, 28, 35, 48, 49, 66, 90, 94
World Heritage in Danger 22–23, 27,
    100
World Heritage institutions 2, 15, 22,
    26, 28, 46, 56, 62, 67, 69–72, 77,
    92, 94, 98, 102–105, 111, 115,
    118–119
World Heritage List 20, 23, 30, 35,
    48–50, 57, 66, 90, 100–102, 105,
    107, 118–119, 136–138
World Heritage sites 20, 22–23, 28,
    49–**50**, 51, 53, 65–66, 71, 92, 94,
    98, 104–105, 110–111, 115
World Heritage Tentative List of
    UNESCO 56
World War II 16, 20, 95

Yanagi, Muneyoshi 82, 86
Yanagita, Kunio 11, 20n11
Yasukuni Shrine 42, 44, 47n3
Yunotsu 62
Yushukan 42–44

zero vector 118–119, 139n7; situation
    119, 127